Macmillan/McGraw-Hill Science

Electrical Energy

AUTHORS
Mary Atwater
The University of Georgia
Prentice Baptiste
University of Georgia
Lucy Daniel
Rutherford County Schools
Jay Hackett
University of Northern Colorado
Richard Moyer
University of Michigan, Dearborn
Carol Takemoto
Los Angeles Unified School District
Nancy Wilson
Sacramento Unified School District

High voltage cross country power lines

Macmillan/McGraw-Hill School Publishing Company
New York Chicago Columbus

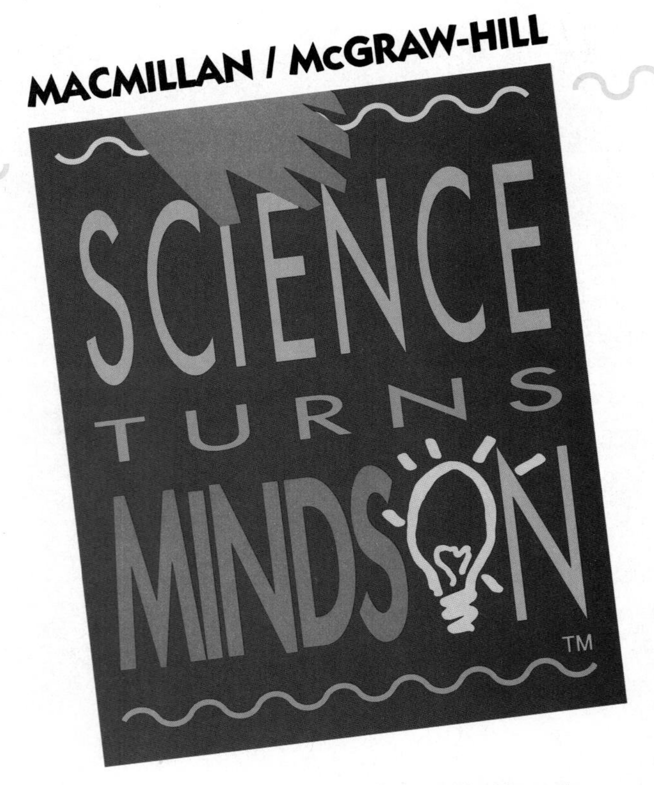

CONSULTANTS

Assessment:

Janice M. Camplin
Curriculum Coordinator, Elementary Science
Mentor, Western New York
Lake Shore Central Schools
Angola, NY

Mary Hamm
Associate Professor
Department of Elementary Education
San Francisco State University
San Francisco, CA

Cognitive Development:

Dr. Elisabeth Charron
Assistant Professor of Science Education
Montana State University
Bozeman, MT

Sue Teele
Director of Education Extension
University of California, Riverside
Riverside, CA

Cooperative Learning:

Harold Pratt
Executive Director of Curriculum
Jefferson County Public Schools
Golden, CO

Earth Science:

Thomas A. Davies
Research Scientist
The University of Texas
Austin, TX

David G. Futch
Associate Professor of Biology
San Diego State University
San Diego, CA

Dr. Shadia Rifai Habbal
Harvard-Smithsonian Center for Astrophysics
Cambridge, MA

Tom Murphree, Ph.D.
Global Systems Studies
Monterey, CA

Suzanne O'Connell
Assistant Professor
Wesleyan University
Middletown, CT

Environmental Education:

Cheryl Charles, Ph.D.
Executive Director
Project Wild
Boulder, CO

Gifted:

Sandra N. Kaplan
Associate Director, National/State Leadership
Training Institute on the Gifted/Talented
Ventura County Superintendent of Schools Office
Northridge, CA

Global Education:

M. Eugene Gilliom
Professor of Social Studies and Global Education
The Ohio State University
Columbus, OH

Merry M. Merryfield
Assistant Professor of Social Studies and Global Education
The Ohio State University
Columbus, OH

Intermediate Specialist

Sharon L. Strating
Missouri State Teacher of the Year
Northwest Missouri State University
Marysville, MO

Life Science:

Carl D. Barrentine
Associate Professor of Biology
California State University
Bakersfield, CA

V.L. Holland
Professor and Chair, Biological Sciences Department
California Polytechnic State University
San Luis Obispo, CA

Donald C. Lisowy
Education Specialist
New York, NY

Dan B. Walker
Associate Dean for Science Education and Professor of Biology
San Jose State University
San Jose, CA

Literature:

Dr. Donna E. Norton
Texas A&M University
College Station, TX

Tina Thoburn, Ed.D.
President
Thoburn Educational Enterprises, Inc.
Ligonier, PA

Macmillan/McGraw-Hill School Division
10 Union Square East
New York, New York 10003

Printed in the United States of America

ISBN 0-02-274268-9 / 4

4 5 6 7 8 9 VHJ 99 98 97 96 95 94 93

An electrical energy generating plant

Mathematics:
Martin L. Johnson
Professor, Mathematics Education
University of Maryland at College Park
College Park, MD

Physical Science:
Max Diem, Ph.D.
Professor of Chemistry
City University of New York, Hunter College
New York, NY

Gretchen M. Gillis
Geologist
Maxus Exploration Company
Dallas, TX

Wendell H. Potter
Associate Professor of Physics
Department of Physics
University of California, Davis
Davis, CA

Claudia K. Viehland
Educational Consultant, Chemist
Sigma Chemical Company
St. Louis, MO

Reading:
Jean Wallace Gillet
Reading Teacher
Charlottesville Public Schools
Charlottesville, VA

Charles Temple, Ph. D.
Associate Professor of Education
Hobart and William Smith Colleges
Geneva, NY

Safety:
Janice Sutkus
Program Manager: Education
National Safety Council
Chicago, IL

Science Technology and Society (STS):
William C. Kyle, Jr.
Director, School Mathematics and Science Center
Purdue University
West Lafayette, IN

Social Studies:
Mary A. McFarland
Instructional Coordinator of Social Studies, K-12, and
Director of Staff Development
Parkway School District
St. Louis, MO

Students Acquiring English:
Mrs. Bronwyn G. Frederick, M.A.
Bilingual Teacher
Pomona Unified School District
Pomona, CA

Misconceptions:
Dr. Charles W. Anderson
Michigan State University
East Lansing, MI

Dr. Edward L. Smith
Michigan State University
East Lansing, MI

Multicultural:
Bernard L. Charles
Senior Vice President
Quality Education for Minorities Network
Washington, DC

Cheryl Willis Hudson
Graphic Designer and Publishing Consultant
Part Owner and Publisher, Just Us Books, Inc.
Orange, NJ

Paul B. Janeczko
Poet
Hebron, MA

James R. Murphy
Math Teacher
La Guardia High School
New York, NY

Ramon L. Santiago
Professor of Education and Director of ESL
Lehman College, City University of New York
Bronx, NY

Clifford E. Trafzer
Professor and Chair, Ethnic Studies
University of California, Riverside
Riverside, CA

STUDENT ACTIVITY TESTERS
Jennifer Kildow
Brooke Straub
Cassie Zistl
Betsy McKeown
Seth McLaughlin
Max Berry
Wayne Henderson

FIELD TEST TEACHERS
Sharon Ervin
San Pablo Elementary School
Jacksonville, FL

Michelle Gallaway
Indianapolis Public School #44
Indianapolis, IN

Kathryn Gallman
#7 School
Rochester, NY

Karla McBride
#44 School
Rochester, NY

Diane Pease
Leopold Elementary
Madison, WI

Kathy Perez
Martin Luther King Elementary
Jacksonville, FL

Ralph Stamler
Thoreau School
Madison, WI

Joanne Stern
Hilltop Elementary School
Glen Burnie, MD

Janet Young
Indianapolis Public School #90
Indianapolis, IN

CONTRIBUTING WRITER
Laurel Price-Jones

Electrical Energy

Lessons **Themes**

Activities!

EXPLORE

TRY THIS

Features

Links

CAREERS

SCIENCE TECHNOLOGY AND Society

Departments

Electrical Energy

Electricity! What comes to mind when you read that word? Do you think about your video game? A lightning storm? The lights of a big city at night?

Perhaps you live near huge power lines that carry electricity from a power plant to cities far away. If you flew over Niagara Falls, you would see such wires going out in all directions. They take the energy of falling water, convert it into electricity, and carry it to places like New York City and Toronto.

You use electricity when you turn on the lights. You also use it when the lights go out and you turn on a flashlight.

You use electricity when you listen to the radio in the kitchen. You carry it with you when you take along a pocket radio and listen to it through headphones.

Minds On! When you turn on a flashlight, radio, or fan, what happens? How does this happen? Write down your thoughts in your ***Activity Log*** on page 1.

Electricity is a form of energy. You use it every day, from the time your clock or radio wakes you up until you turn off the lights before going to sleep.

Energy is needed to make things move. Without energy, everything in the world would come to a stop. Living things use the energy they get from food. You change that energy into running, thinking, and helping around the home. One form of energy can be changed into another. In a light bulb, electrical energy is changed into light. In your toaster, electrical energy becomes heat.

Long ago, people used only their own energy to make things move. Today, there are other forms of energy people can use to do things for them. Electricity runs the washing machine. Gasoline engines power cars to get us where we want to go. We don't have to walk everywhere. Gas, oil, or electric furnaces keep us warm in the winter. We don't have to use a fireplace and work hard to keep a woodshed filled. Our ability to use many forms of energy to help us with our work can make our lives much easier.

Washing machine

Washboard

Energy To Do Things in the Past

Social Studies Link

Since earliest times, people have depended on their own muscles and the fires they build for energy. Later, they trained animals to help them. The animals plowed fields and pulled wagons. In time, people discovered that the energy in the blowing wind could do work. Sailboats made it easier for people to go far out onto the oceans because they didn't depend on rowing for power anymore.

In the 1700s in Europe people began to use steam power. They burned coal to make steam and then used the steam to power engines. This made it easier and cheaper for people to manufacture goods. Electrical energy was not widely used until the 1900s. Write a story that describes what life was like before electricity. How do you think life then was different from life today? How was it the same?

In this unit you will learn more about how electrical energy works. You will experiment with electricity and learn how to use it safely. You may make electricity part of a new hobby or the start of a career!

Water mills used the energy of flowing water to grind grain.

Dam and electricity generating station

Science in Literature

Literature Link

In these books, you can find ideas for things you can make that use electricity or magnets. You can also learn about how electrical energy is part of the things around you. While you read, look for the many things that electrical energy does.

The Calvin Nullifier
by Gene DeWeese.
New York: Dell Publishing, 1989.

Calvin Willeford and an alien named Dandelion discover the cause of damage to the circuits on a space probe. They must evade an angry sheriff and convince a scientist to help them. When they get the information and tools they need, they travel through space to Uranus to fix the probe. See if you can tell the facts about electricity from the fiction when you read this book.

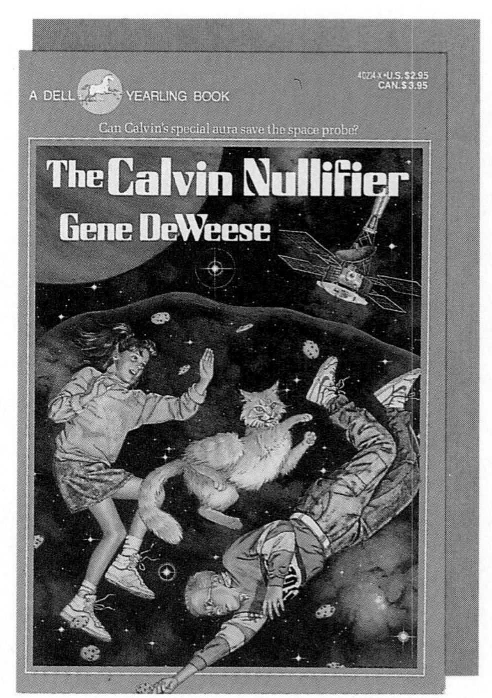

Other Good Books To Read

Benjamin Franklin by Eryl Davies. London: Wayland Publishers, 1981.

Black-and-white historical illustrations enhance this biography of an "experimenter extraordinary." It includes his inquiries into electricity, lightning, oceanography, static electricity, and magnetism.

Electricity and Magnetism by Peter Lafferty, illustrated by Jeremy Grower. New York: Marshall Cavendish, 1989.

Text and experiments provide information about the world of electricity and magnetism. Included are notes about famous scientists: Faraday, Galvani, Oersted, Thales, and Watt.

Electricity and Magnets by Barbara Taylor. New York: Franklin Watts, 1990.

These fun activities can show you more about how electricity and magnetism work. Most of them use materials you can find at home.

Energy All Around Us by Donna Bailey. Austin, TX: Stek-Vaughn, 1991.

This book discusses different kinds of energy, including electricity. It tells where energy comes from, how it is stored and transported, and how it is used.

Energy and Light by Peter Lafferty. New York: Gloucester Press, 1989.

This book defines matter and then explores ways matter can be changed from one form to another through the use of energy. It includes good coverage of all kinds of energy, including electricity and magnetism.

You'll Get a Charge

What makes socks stick together when they come out of a clothes dryer? The energy that makes the clothes cling to each other is like something helpful you use every day.

Electricity is invisible. You can't see it or hear it. But you can see and hear the effects of electricity all around you. Sometimes you can feel the effects of electricity.

Do you ever hear clothes crackle and see them cling to each other when you take them out of the dryer? What you see and hear the clothes do is caused by electricity.

Electricity is important to the way we live. It is used to run refrigerators and light bulbs as well as to run televisions and electric guitars. People have experienced the effects of electricity for many thousands of years. It is only in the last 150 years, however, that people have begun to really understand how electricity works.

Minds On! Think about the cling and crackle of clothes fresh from the dryer. Now, draw a picture in your ***Activity Log*** on page 2 or tell one of your classmates how you think the electricity gets on the clothes while they're in the dryer. ●

Out of This!

EXPLORE Activity!

Sticky Cling

You're going to see one way things get cling, like your clothes in the dryer. You'll also see that there's more than just cling involved.

What You Need

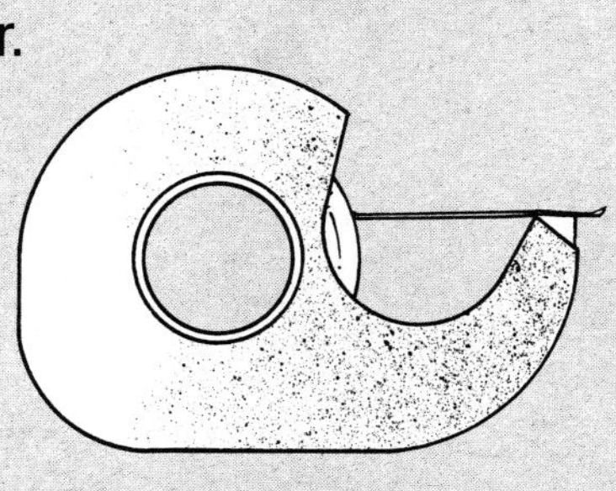
transparent tape

***Activity Log* pages 3-4**

What To Do

1 Tear two strips of tape about 10 cm long off the tape dispenser.

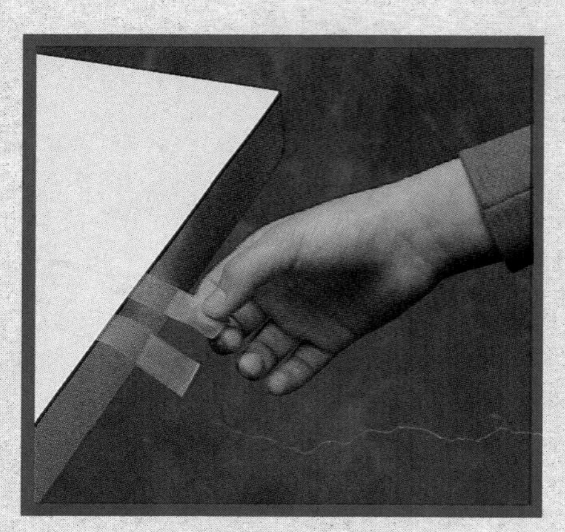

2 Stick them to your desk top, leaving about 1 cm over the edge. Fold the edge back so that there's a nonstick part to grab.

3 Peel the strips off your desk, one at a time, so that the tape doesn't curl up.

4 Now, hold the two strips by their ends and bring them close to each other. What happens? Write your observations in your ***Activity Log.***

5 Have your partner stroke both of the strips several times with his or her fingers. Bring the strips together again. What happens? Write your observations in your ***Activity Log.***

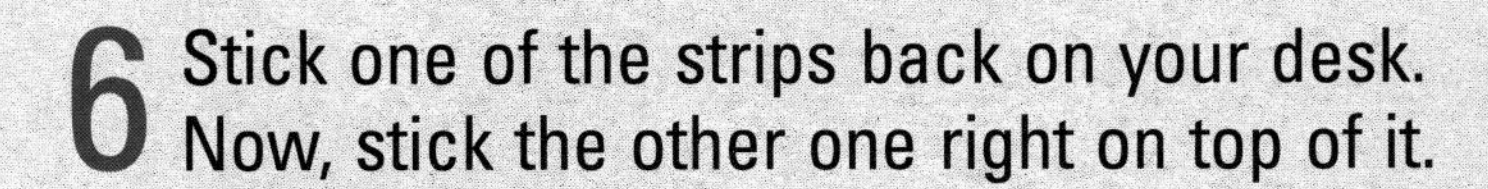

6 Stick one of the strips back on your desk. Now, stick the other one right on top of it.

7 Peel both strips off your desk. Now, peel the two strips apart. Predict what will happen when you bring the two strips together. Have your partner write your prediction in your ***Activity Log.***

8 Test your prediction and then write your observations in your ***Activity Log.***

9 Have your partner stroke the strips several times again. Predict what will happen when you bring the two strips together after they've been stroked.

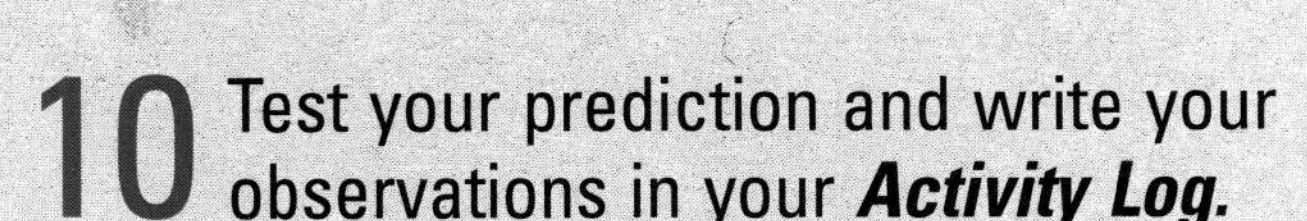

10 Test your prediction and write your observations in your ***Activity Log.***

What Happened?

1. What happened when you brought the strips of tape near each other the first time? The second time?
2. What happened after your partner stroked the strips?

What Now?

1. Why do you think you got different results depending on what you did to the tape?
2. What other things have you seen that behave as the tape does? How were they like the tape?

Electric Charge

In the Explore Activity, you pulled two pieces of tape off your desk and found that the pieces pushed away from each other. When you pulled both pieces off your desk, then off each other, they pulled toward each other. After your partner stroked the pieces of tape, neither piece pushed or pulled.

The tape pieces did two unusual things because of electricity. How can two things be caused by only one thing? The strips of tape moved because they had an electrical charge. The tape did two unusual things because there are two opposite kinds of electrical charge, positive (pos´ i tiv) (+) and negative (neg´ ə tiv) (–). These two types of electrical charge interact.

About 2,600 years ago, around 600 B.C., the Greeks noticed the same sort of thing you did. When pieces of a yellow stone they called *elektron* (i lek´ tron) were rubbed with sheep's wool, they did a strange thing—they pulled other small objects to them.

In A.D. 1570, William Gilbert, a British doctor, discovered he could make amber beads push away from each other, and make a glass bead and an amber bead pull toward each other. He called what he saw electricity, *after* elektron, *the Greek word meaning "amber."*

"Elektrons" play an important part in electricity today, too. You may have learned that all matter and all living things are made of very, very tiny pieces called *atoms*. An atom is the smallest piece of an element that could still be identified as that element. Atoms are made of even smaller pieces.

Atoms usually have the same number of protons and electrons. The amount of positive (+) and negative (–) electrical charge is the same. When the pieces of tape didn't pull toward or push away from each other, the atoms in the tape had the same amount of (+) and (–) electrical charge. In some materials, electrons are easily pulled off atoms. This means that the electrons (–) can move to another material.

Illustration not to scale

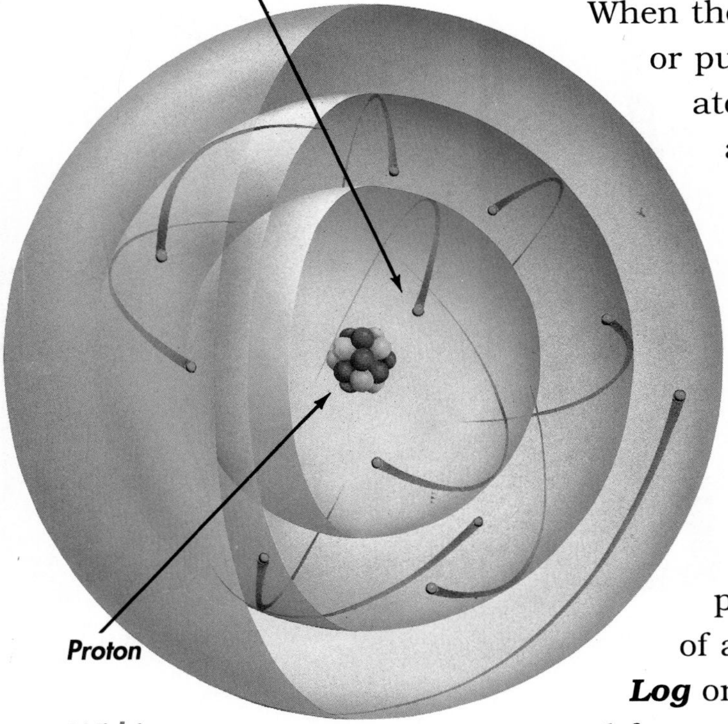

Electron

Proton

Within each atom there are even tinier parts, each of which has the smallest possible amount of electricity, one ***charge. Electrons*** *(i lek´ tronz) each have one negative (–) charge.* ***Protons*** *each have one positive (+) charge.*

Minds On! You can see where electrons move. Use a pencil to draw a picture of your desk and a picture of a piece of tape in your ***Activity Log*** on page 5. Put five positive (+) signs and five negative (–) signs on each picture. Imagine that you have pulled some of the electrons (–) off the desk and onto the tape. Erase three of the (–) signs on the desk and add them to the tape. Does the tape have more positive or more negative signs? Does the desk have more positive or more negative signs?●

In the Explore Activity, both of the pieces of tape pulled electrons off your desk. They both had more negative electrical charge and they pushed away from each other because like charges repel.

When you pulled the two stuck-together pieces of tape off your desk, they had extra electrons. Then, when you pulled them apart, the sticky side pulled most of the extra electrons off the bottom piece. The top piece then had more extra negative electrical charge. The bottom piece had many fewer electrons than it started with, so it had more positive electrical charge. Since one piece had more negative electrical charge and one piece had more positive electrical charge, they pulled toward each other.

Each balloon has a different static electrical charge, one is positive and one negative. The different charges pull toward each other. Do the glass and amber beads on page 16 have like or different charges?

The kind of electricity the pieces of tape had is called **static electricity** (stat´ ik i lek tris´ i tē). It's called *static* because the electrical charge is on something and not moving (even though it moved to get there). You've probably seen the cling that clothes have when they come out of the dryer. It's called *static cling* because it is caused by static electricity. Like the charges on the pieces of tape, it comes from different kinds of material rubbing against each other.

A ***static electrical charge*** *is an increased concentration of electrical charge in one place. Each balloon in the picture has a negative electrical charge, because each has more electrons than protons. The like charges push away from each other.*

On the Wall — Math Link

Blow up a balloon and tie it closed. Rub it against your hair or your clothes for a minute and stick it on the wall. Does it stay? Use page 6 of your ***Activity Log*** for your answers. What makes it stick on the wall? How long can you make the balloon stay up? Use the second hand on your classroom clock to time your balloon. Start keeping time when you place the balloon on the wall. What makes the balloon stay up longer?

Does it make a difference if you rub the balloon 30 seconds? 90 seconds? 120 seconds? Make a bar graph of your results. Rub the balloon again. Now have your partner rub all over the balloon with his or her hands. Does the balloon still stick to the wall? What changed?

When you rub a balloon on your hair, you give it more of one type of charge. When you rub a balloon with your hands, the extra charge moves to your body, spreads out, and travels to the ground. Getting rid of extra electric charge by safely transferring it to the ground is called **grounding** (ground´ ing) the charge. Grounding is the easiest way to get rid of static electricity.

Effects of Static Electricity

The small amounts of static electricity you've experienced can cause an annoying shock. Big amounts of static electricity—lightning—can give a very dangerous shock that can kill or start fires. Static electricity builds up in clouds when the atoms bump and rub against each other. When that electricity travels to the ground, you see lightning.

A lightning rod is put on the roof of a building to carry the electricity away from the building. Part of the reason this works is that the rod is connected to the ground with a wire.

A lightning rod grounds the enormous static electrical charge of a lightning bolt.

TRY THIS **Activity!**

Clean Up!

What You Need

pepper, balloon, small piece of wool cloth, *Activity Log* page 7

Pretend that you just spilled some pepper at home. You need to clean it up and you don't have a vacuum cleaner. In your ***Activity Log,*** draw how you could pick up the pepper with the materials you have. Then, explain to one of your classmates why your method works. Try it.

Small amounts of static can do useful things, but they can be dangerous to electronic machines.

Computer Technician

Computer technicians help keep computers working. They clean computers and replace or repair broken parts of computer systems. They need to know about computers and the equipment computers use.

Computers cannot withstand big, sudden changes in electric current. If a static charge flowed through a computer circuit (or chip), it could destroy the chip. People can become electrically charged when they walk on carpet or when different clothes rub together. To protect against this, computer technicians frequently wear a special kind of watchband that grounds static electricity.

To become a computer technician, you will need to become as familiar as you can with computers. Take any class in school that offers time on a computer and take as many math courses as you can.

Sum It Up

You probably think of the electrical energy you use at home as something that does work for you and makes things move. Static electricity is a form of energy, too. You've seen that it can make things move. Understanding the electrical energy you use every day is easier when you understand that static electricity is caused by electrons that move from one place to another.

Critical Thinking

1. To put a charge on a balloon, you rub it on your sweater. If the balloon has a negative charge, what charge does your sweater have? How do you know?

2. Carpet manufacturers have spent time and money to find chemicals that will keep carpet from making static electricity. Why do you think they did this?

3. A friend shows you a balloon with a mystery charge on it. How can you find out if it is a positive, a negative, or a neutral (balanced) charge?

Current Events

How does the energy get to the lights in your home? Controlling the flow of electricity is the key.

Static electricity is fun to play with, but it would be difficult to use it to light our homes or run machinery. We'd have to rub things or pull things off each other, then carry them to where we needed to use them. Besides that, it's hard to make lots of static electricity. How do you think we do get electricity from one place to another? What happens if the electricity doesn't get to where we need it?

Early in the evening on November 9, 1965, all the lights in New York City went out. All the refrigerators stopped, electric fans quit, and radios were quiet. The loss of electricity was called a *blackout.* For several hours there was no electricity in a city of millions of people. Usually bright and noisy, New York City was suddenly dark and quiet.

1965 blackout in New York City

A power outage

Minds On! What would happen if your town were without electricity for 24 hours? In your ***Activity Log*** on page 8, make a list of reasons why it would be fun, why it would be dangerous, and why it would be inconvenient.

EXPLORE

Activity!

Pathmaking

Understanding a very simple thing like making a bulb light will help you understand how the lights in your home work. In this activity you will see how to control where electricity goes.

What You Need

D-cell

Activity Log pages 9-10

flashlight bulb

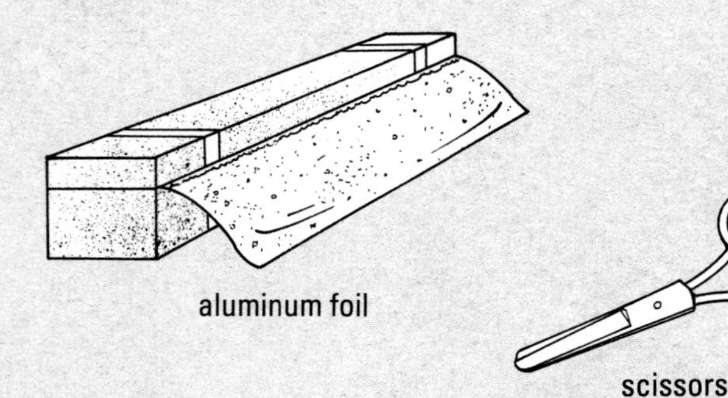

aluminum foil

scissors

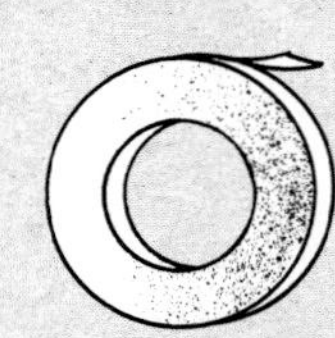

about 20 cm of masking tape

What To Do

1. First, make a wire by sticking the tape onto the aluminum foil.

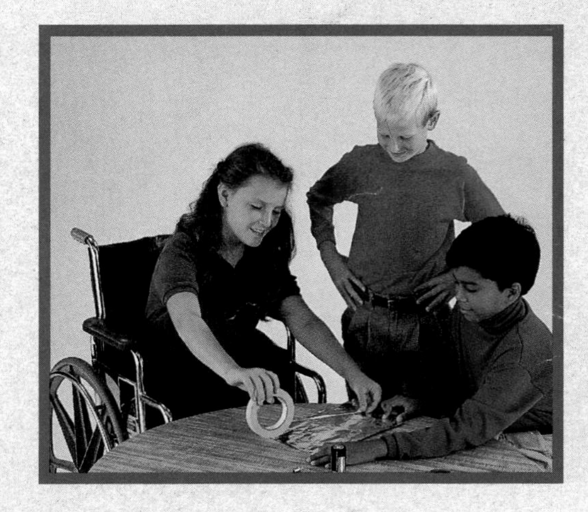

2 Cut around the tape and fold it in half lengthwise with the foil side out.

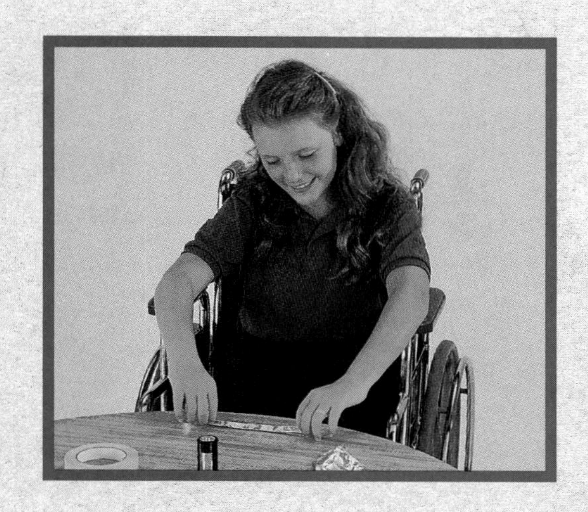

3 Now, find as many ways as you can to make the flashlight bulb light, using your wire and the D-cell. Draw in your ***Activity Log*** each arrangement you try—even if it doesn't work. Be sure to mark the ones that do work.

What Happened?

1. How did you make the bulb light each time?
2. What happened if you used the masking tape side of your wire?

What Now?

1. What was the same about each of the ways you lit the bulb?
2. If you owned a flashlight and it did not light, what would you check to find out why it wasn't working?

Circuits

In the Explore Activity, you probably found more than one way to make the bulb light. They all had one thing in common, though. The wire had to be connected so that you could trace a line from one end of the D-cell, through the bulb, and back to the other end of the D-cell.

Look back at the drawings you made in your ***Activity Log*** on page 9 during the Explore Activity. Try to trace a path from one end of the D-cell to the other, going through the light bulb. Can you follow an unbroken path on all the ways that worked?

The train track is a complete circle pathway for a train to follow. A circuit is a complete circle pathway for electricity to follow.

Electricity explains why the light bulb lit. Electricity has to have something to move through to get to the light bulb. A path that electricity can move through is called a **circuit** (sûr´ kit).

Static electricity is extra electrical charge that stays in one place. **Current electricity** (kûr´ ənt i lek tris´ i tē) is moving electrical charge. It's called *current electricity* because the electrical charge moves somewhat like the currents in a river or stream. The current electricity we use to do things every day is moving electrons. A D-cell like the one you used in the Explore Activity is a way to make an electric current flow.

TRY THIS

Activity!

Water Wire

What You Need

1-m piece of clear plastic tubing, food coloring, plastic box, funnel, water, *Activity Log* page 11

How is water moving in a tube like electrons in a wire? Form the tubing into a shallow U-shape with one end about 5 cm higher than the other. Have your partner hold the low end so that it's over the plastic box. Using the funnel, fill the tube with water. Add a small drop of food coloring to the water at the high end of the tube. Add more water. How does the food coloring move through the tube? What would happen if you cut the tube in the middle?

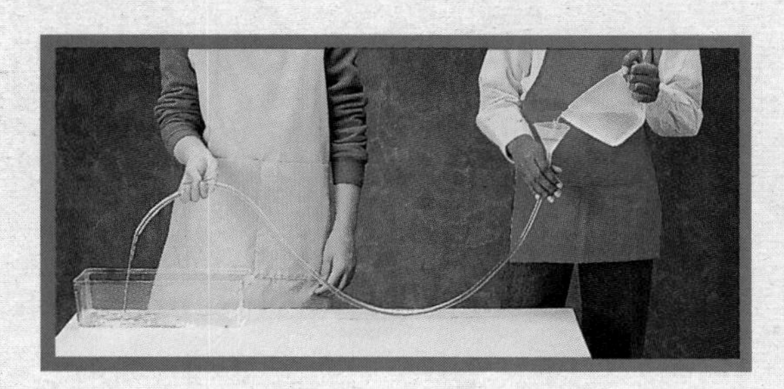

In the Try This Activity, the food coloring moved almost the way a bunch of electrons move in a wire. Although the coloring started in a clump, it spread out and some of it got to the end of the tube before the rest. This model, like all models, has some flaws. The water doesn't move nearly as fast as electrons do.

If you cut the tube in the middle, the water wouldn't get to the other end. When you had the wire touching correctly in the Explore Activity, you got electricity to the other end and the bulb lit. The wire made a **closed circuit** because it had no breaks or places where the current couldn't flow.

If you had cut the wire or let it come off one end of the D-cell, the bulb would have gone out. A circuit with a break in it is called an **open circuit.**

The train is like an electric current. It is moving around the track circuit. Like electricity, it has energy because it is moving.

TRY THIS **Activity!**

What Turns It On?

What You Need

working circuit from the Explore Activity, extra foil wire, piece of cardboard about 6 cm x 6 cm, two paper brads, *Activity Log* page 12

Attach one of the wires to the bottom of the D-cell. Use a pen cap to make two holes in the cardboard about 4 cm apart. Attach both wires to the cardboard with brads, using the holes. Make sure the metal of the wire touches the brads. Make two lists in your ***Activity Log***—one of things that will connect the brads and make the bulb light, and one of things that will not. Try to close the circuit across the brads using the items on your lists. Mark the items that work. Can you use them to make an easy way to turn the bulb on and off?

Try closing the circuit with your finger. What happens? Tell one of your classmates about your results, especially any that surprised you.

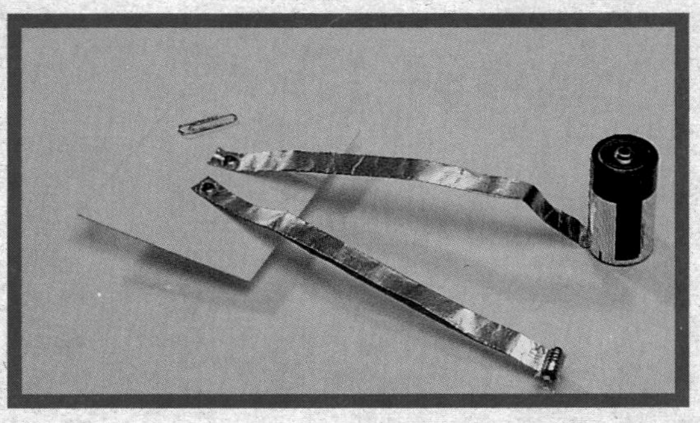

Like the space between the brads in the Try This Activity, there is a break in a circuit that can be opened or closed. The switch is open, and the train is stopped because the electric current is stopped.

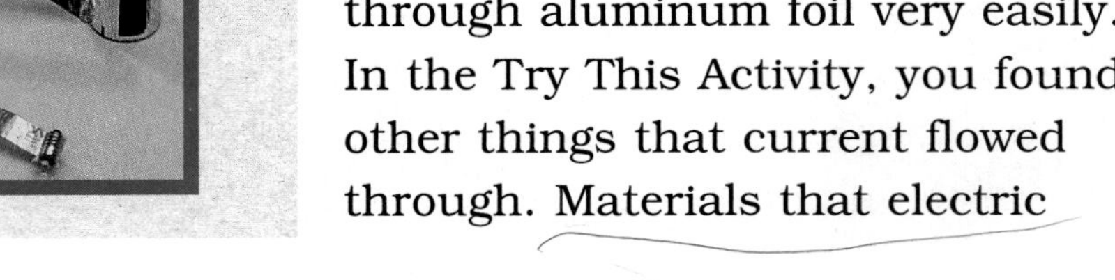

Conductors and Insulators

Some materials will close a circuit. Electrical current flows through aluminum foil very easily. In the Try This Activity, you found other things that current flowed through. Materials that electric

The girl has opened the train circuit by "cutting" the track. The train couldn't move around the track. Current can't flow when a circuit is open.

current flows through well are called **conductors** (kən duk´ tərz). Most metals are good conductors. Your wire works because aluminum is a metal that conducts well. Other common things, like water, also conduct.

Electric current doesn't flow well through all materials. If you try to hold the masking tape side of your wire against the D-cell, your circuit won't work. Masking tape doesn't have electrons available to conduct electricity. Materials that electric current doesn't flow through well are called **insulators** (in´ sə lā´ tərz). Most plastic and rubber materials are insulators.

It's dangerous to stick your fingers (or anything else!) into electrical sockets. In part, this is because you conduct electricity since your body is made mostly of water. Electricity can flow through your body and burn you. Household current can be dangerous if people touch bare wires. All the wires around your home have an insulator around the conductor. Because the insulator can be scraped or damaged, household wiring is put inside the walls.

Minds On! In your ***Activity Log*** on page 13, list or draw some other ways in which wires are kept out of the way for safety.●

Sometimes we need to break conductors and open circuits, as you do when you turn off the lights or stop an electric train. The easiest way to do this is to use a switch. A **switch** is a device that is used to open and close circuits without unscrewing bulbs or disconnecting wires. The brads in the Try This Activity were part of a switch that you opened and closed with various materials.

Some materials don't conduct electric current as well as copper and aluminum. When electrical current flows through these materials, they get hot. They are called **resistors** (ri zis´ tərz) because they resist the flow of electric current. The light bulb in your circuit is a resistor. When current flows, the wire inside the bulb gets very hot and you see light. This is why light bulbs get hot after they've been on for a while. This heat isn't always good. In the hot summer time, working bulbs can make a room much warmer.

The electrical current comes into the bulb from the metal side, flows through the filament, and out the tip. Look at the bulb you used in the Explore. Find these parts in it.

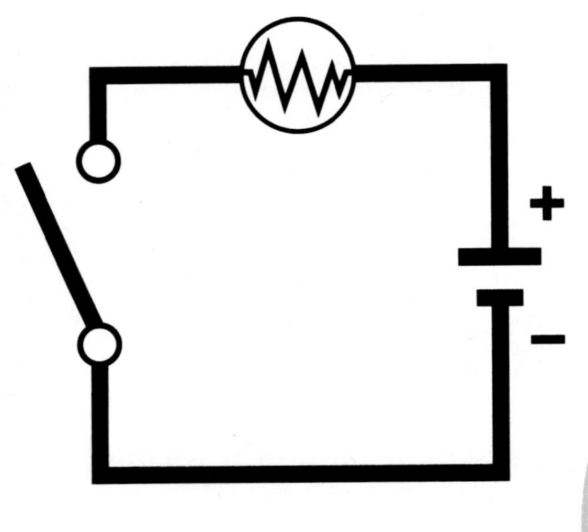

People don't always have the time to draw pictures of wires, cells, and bulbs to show what's in a circuit and how it works. Instead, they use symbols to represent the parts of the circuit. Can you match the cell, wire, light bulb, and switch symbols with the parts of the circuit shown in the picture?

Simple drawings with symbols and lines to show the circuit parts are called **circuit diagrams** (sûr´ kit dī´ ə gramz´). People who build and design circuits use diagrams to communicate ideas so that they don't have to sketch complicated pictures.

The Calvin Nullifier Literature Link

When the story *The Calvin Nullifier* begins, the household current at Calvin's house flickers and goes out. Read pages 14 and 15 to find out how this electricity change is connected to the nullifier and an alien from space. You can also read more about circuit diagrams and how they help us design the things that use electricity.

Lewis H. Latimer

Today's light bulb filaments are made of metal. In the past, light bulb filaments were made of thin, fragile pieces of carbon. Many of those carbon filaments were manufactured by a process Lewis H. Latimer developed.

Making the Best Use of Electricity

Did you notice the label on your D-cell that said *1.5V?* That means 1.5 volts. A **volt** (vōlt) is a unit for measuring the force that makes electrons move in an electric current. The cells you use are quite safe because they have such a low voltage. Household current is at least 110 volts. That's almost 100 times more. That's one reason why household current is so dangerous.

Sometimes we use the energy of household current to recharge special kinds of cells. Even though rechargeable (rē charj´ ə bəl) cells are more expensive to buy, they are usually cheaper to use. The energy in the rechargeables is clean and nonpolluting, so many of them together are used to run small equipment that would otherwise run on gasoline or other polluting fuels.

Focus on Technology

Electric Cars

Many large cities like Los Angeles and Mexico City have lots of air pollution from gasoline car exhaust. These cities are interested in finding ways to avoid this source of pollution. People can walk or use public transportation, such as buses and subways, to reduce pollution. Electric cars may also be a solution. They don't produce any pollution where they're used. They use rechargeable batteries instead of gasoline. The batteries are recharged from household electricity. However, the cars and batteries cost a lot of money to make and the cars won't travel very far on one fill-up of electricity. Even so, cities with big air pollution problems are considering requiring some use of these cars in the future. Do you think they should?

Electric car

Rechargeable batteries in electric car

Sum It Up

Now you know one big difference between static electricity and current electricity—it's easier to get current electricity to do things for us. Your circuits made light in much the same way that your home is lighted by electricity. Many of the things you use every day to make light, heat, or movement are run by electricity. Each piece of electrical equipment, from toasters to electric cars, has a circuit inside it. Sometimes the circuit is very simple. Sometimes the circuit is very complicated, containing many parts that control the flow of electricity.

Current electricity is a useful form of energy because it is easy to send it where we need it by using the way it interacts with circuits made with conductors, insulators, and resistors.

Critical Thinking

1. Using what you know about conductors and insulators, why do you think it is dangerous to have hair dryers, radios, or appliances in and around a bath or sink that's full of water?

2. Why is it dangerous to have cracked or peeling electric cords that leave wires exposed?

3. A resistor in an electrical circuit changes electrical energy to another form of energy, heat or light. What might a resistor in a water wire circuit change the water's energy to?

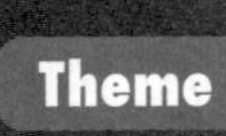

DOWN

There's usually more than one way to get something done. That's also true of electrical circuits. There are many ways to connect light bulbs and turn them on and off.

You've made a circuit with just one light bulb and one D-cell. Sometimes you want to turn on more than one light bulb with the same switch. Or maybe you want to light a bulb with more than one D-cell. How do you think that's done? Have you ever seen all the lights on a sports field come on at once? How does that happen?

You can't see them, but there are wires running to all the wall sockets in your home and school. If each socket has its own circuit, why do several lights and appliances usually go out all at once if there's a problem? Shouldn't only one appliance or light with a problem go out? How are all the lights and appliances connected?

Minds On! Which light switches at home or at school turn on more than one light? In your ***Activity Log*** on page 14, draw a picture of how you think these lights are connected.

EXPLORE Activity!

Different Paths

You're going to build four complex circuits using circuit diagrams. These circuits will give you some idea of how the wiring in your own home is set up. They'll also show you more about how electrical energy interacts with the parts of a circuit. *Safety Tip:* You should never experiment with the household current from any source! It is very dangerous. The D-cells you are using are a safe source of electricity for you to use.

What You Need

2 bulb holders

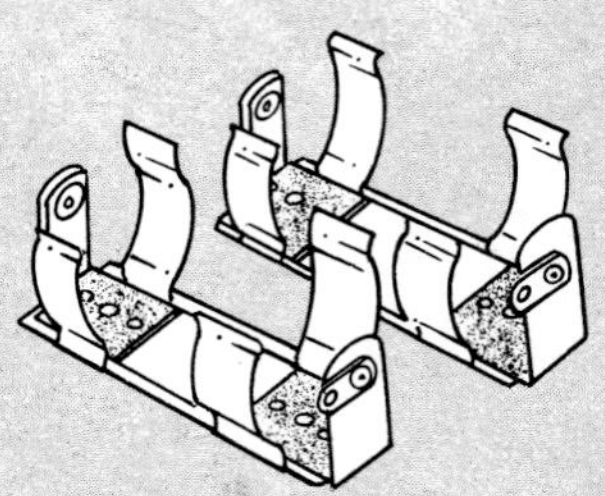

2 D-cell holders

2 light bulbs

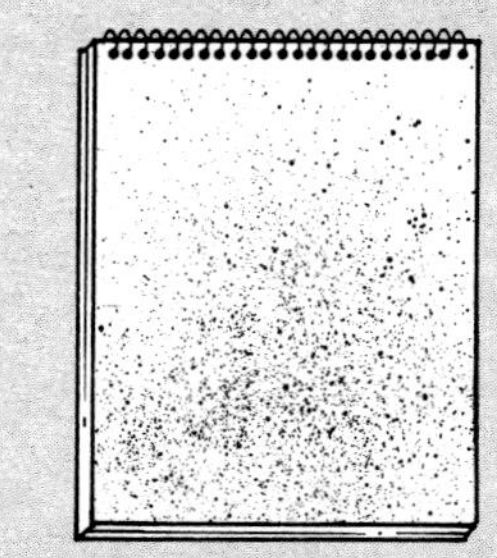

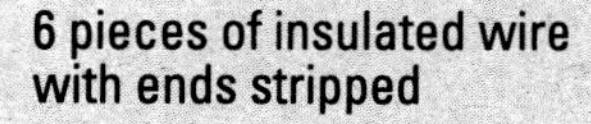

6 pieces of insulated wire with ends stripped

2 D-cells

Activity Log pages 15-16

What To Do

1 Build circuit 1 using the diagram shown.

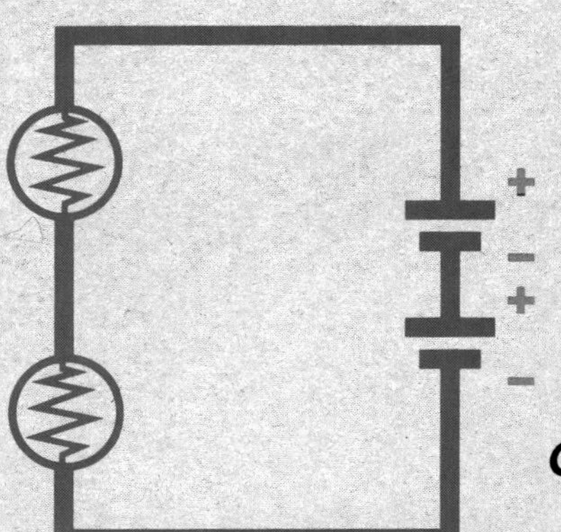

Circuit 1

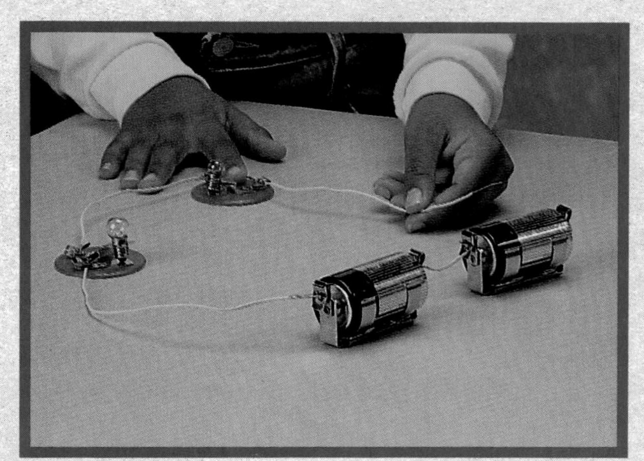

See the *Safety Tip* in the introduction.

2 When you have the circuit complete and working (both bulbs lit), predict what will happen when you remove one of the bulbs. Test your prediction.

3 Predict what will happen if you disconnect one of the D-cells. Test your prediction.

4 Repeat steps 1–3 for the other three circuits shown. Build circuit 3 last. When you have finished, just disconnect it. You will also need it for the Try This Activity on page 40.

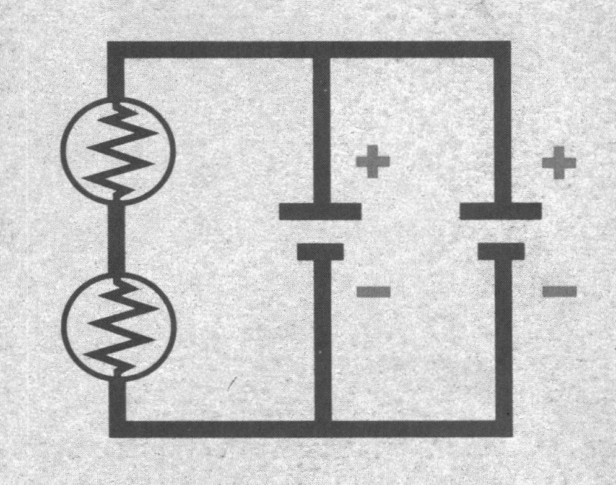

Circuit 2

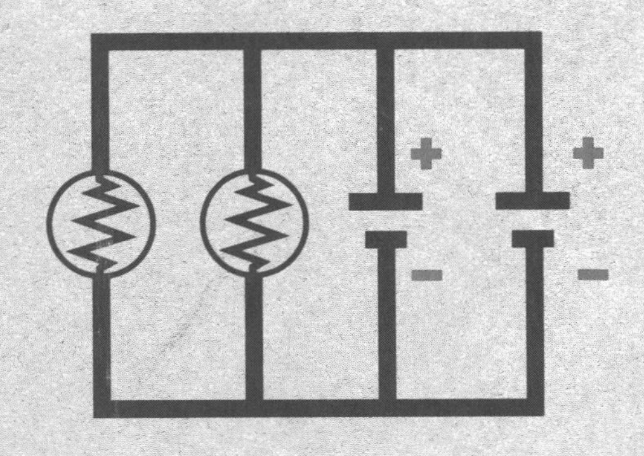

Circuit 3

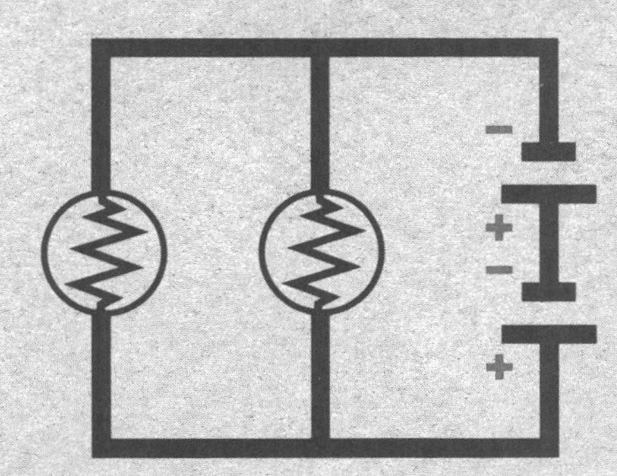

Circuit 4

What Happened?

1. What happened when you removed a light bulb in each of the circuits?
2. What happened when you disconnected one of the D-cells in each of the circuits?

What Now?

1. Why did the other bulb go out in some circuits but not in others when you removed a bulb?
2. Why did both bulbs go out in some circuits but not in others when you disconnected a D-cell?
3. How can removing and replacing a bulb be like opening and closing a switch?
4. Which kind of circuit do you think would work best to light your home?

Kinds of Circuits

The D-cells and light bulbs in your circuit were connected in two different ways. That's why the bulb only went out sometimes when you disconnected part of the circuit. How the D-cells and bulbs are connected also controls how much electricity flows through the bulbs. That's why in some circuits the bulbs were dimmer than in others. The two different ways the bulbs were connected are called *in series* and *in parallel.*

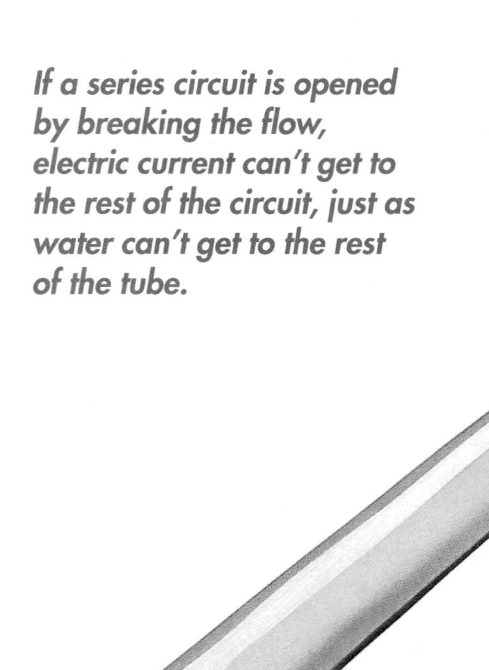

If a series circuit is opened by breaking the flow, electric current can't get to the rest of the circuit, just as water can't get to the rest of the tube.

Minds On! Look back at the circuit diagrams in the Explore Activity on pages 36–37. Use your finger to trace a line from the D-cell through each of the light bulbs. Can you tell why one type of circuit is called *parallel?* Can you tell why one is called *series?*●

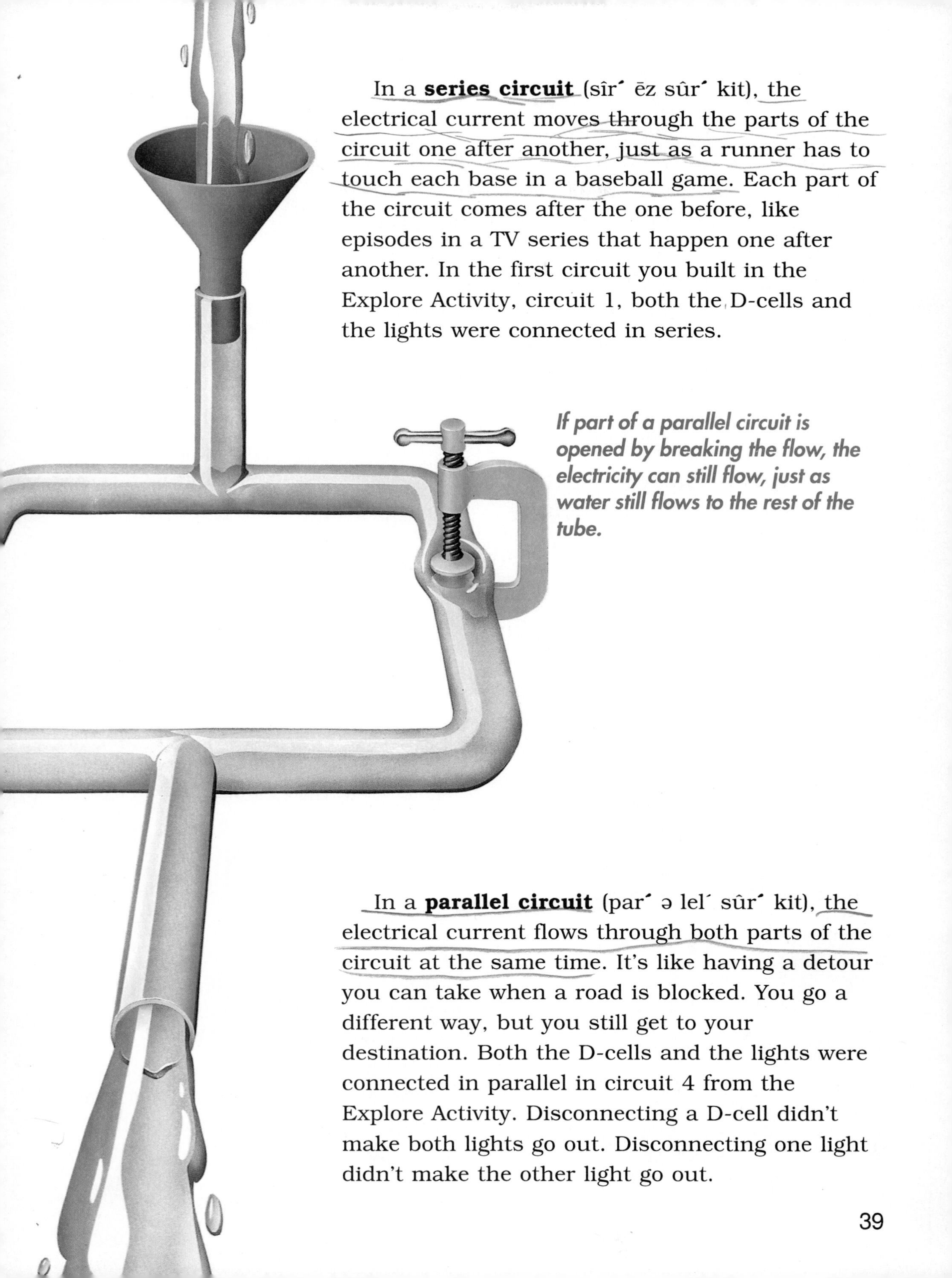

In a **series circuit** (sîr´ ēz sûr´ kit), the electrical current moves through the parts of the circuit one after another, just as a runner has to touch each base in a baseball game. Each part of the circuit comes after the one before, like episodes in a TV series that happen one after another. In the first circuit you built in the Explore Activity, circuit 1, both the D-cells and the lights were connected in series.

If part of a parallel circuit is opened by breaking the flow, the electricity can still flow, just as water still flows to the rest of the tube.

In a **parallel circuit** (par´ ə lel´ sûr´ kit), the electrical current flows through both parts of the circuit at the same time. It's like having a detour you can take when a road is blocked. You go a different way, but you still get to your destination. Both the D-cells and the lights were connected in parallel in circuit 4 from the Explore Activity. Disconnecting a D-cell didn't make both lights go out. Disconnecting one light didn't make the other light go out.

Controlling Circuits at Home

The light bulbs in your home are connected in parallel, and now you know one reason why. If one bulb burned out, you wouldn't want all the lights to go out. Did you notice that the bulbs in the Explore Activity were dimmer when they were connected in series? This is another reason why your home lights are connected in parallel. If they weren't, the lights on the circuit would probably be so dim you couldn't see to do anything by them.

Some things have to be connected in series. If switches were connected in parallel with the lights they control, you couldn't easily turn off the lights. Remember what happened when you took one of the bulbs out in the Explore Activity.

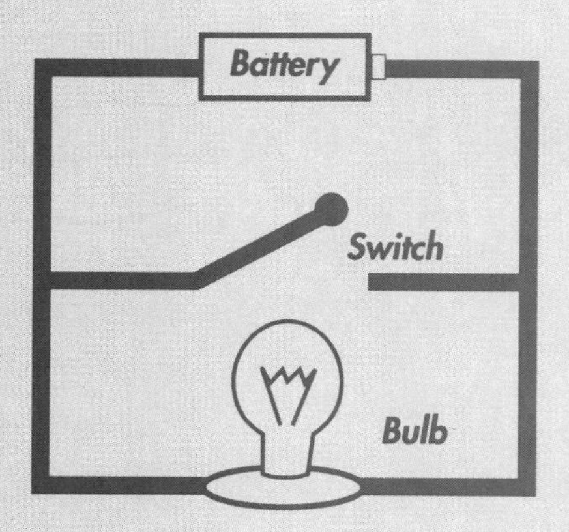

Now, look at the picture showing a switch in place of a bulb. When the bulbs were connected in parallel, removing one bulb didn't make the other go out. The open switch doesn't make the light go out. Switches must be connected in series with what they control.

TRY THIS **Activity!**

Switch Better

What You Need

circuit 3 from the Explore Activity, paper clip switch, extra wire, *Activity Log* page 17

Can you control one light in a parallel circuit? What would happen if you opened the switch in the circuit shown? Build the circuit with a paper clip switch and test it. Describe in your ***Activity Log*** how this works like the lights you use at home.

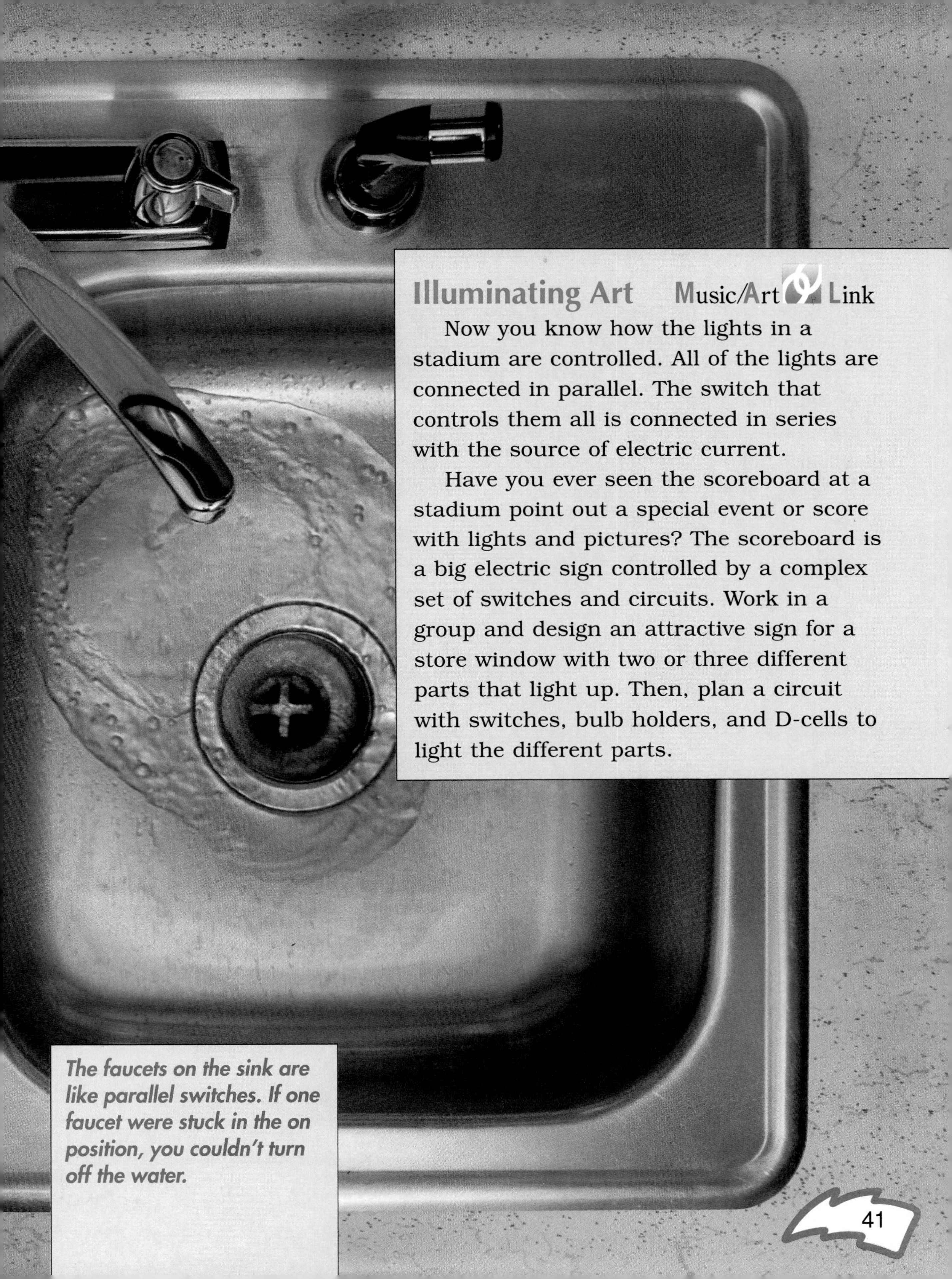

Illuminating Art

Music/Art Link

Now you know how the lights in a stadium are controlled. All of the lights are connected in parallel. The switch that controls them all is connected in series with the source of electric current.

Have you ever seen the scoreboard at a stadium point out a special event or score with lights and pictures? The scoreboard is a big electric sign controlled by a complex set of switches and circuits. Work in a group and design an attractive sign for a store window with two or three different parts that light up. Then, plan a circuit with switches, bulb holders, and D-cells to light the different parts.

The faucets on the sink are like parallel switches. If one faucet were stuck in the on position, you couldn't turn off the water.

When Circuits Don't Work

Have you ever had only some of the lights or appliances in your home go out? Your parents probably had to replace a fuse or reset a circuit breaker. **Fuses** (füz´ əz) and **circuit breakers** (sûr´ kit brāk´ ərz) are devices that are used to keep too much electrical current from flowing through wires. Too much current is very dangerous because it can heat the wires and start fires where you can't see them—inside your walls. This is a problem for the portable home shown in the picture, too.

Circuit breakers are special switches. When too much electricity flows through a circuit breaker, a piece of metal gets hot and expands to push the switch open. A spring holds the switch open. Circuit breakers can be reset and used again.

TRY THIS **Activity!**

"Short" Stop

What You Need

20 cm of wire, circuit 1 from the Explore Activity, *Activity Log* page 19

What happens if you provide a shorter, easier path for electrical current to follow? Connect one end of the extra wire to a bulb holder where the wire from the D-cells connects. Touch the other end of the wire briefly to the last bulb holder connection.

What happens? How is this like the parallel circuit with a closed switch and a bulb? In your ***Activity Log,*** tell why you think this is called a *short circuit.*

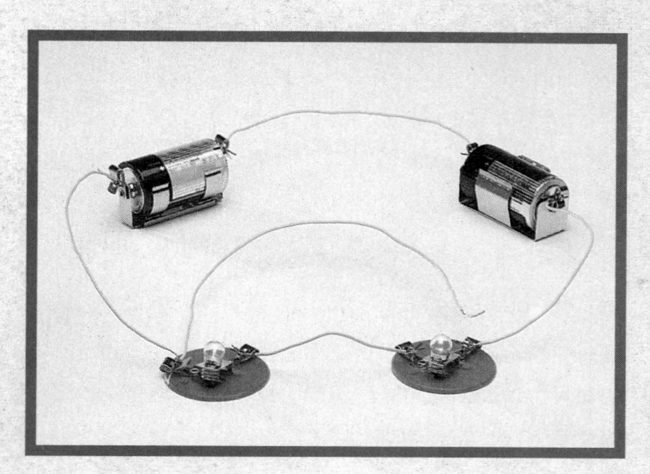

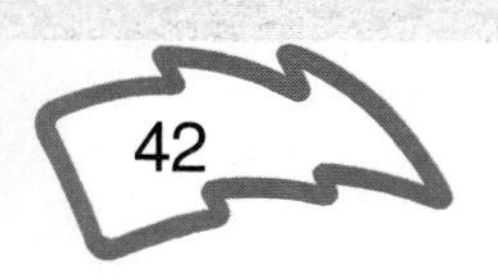

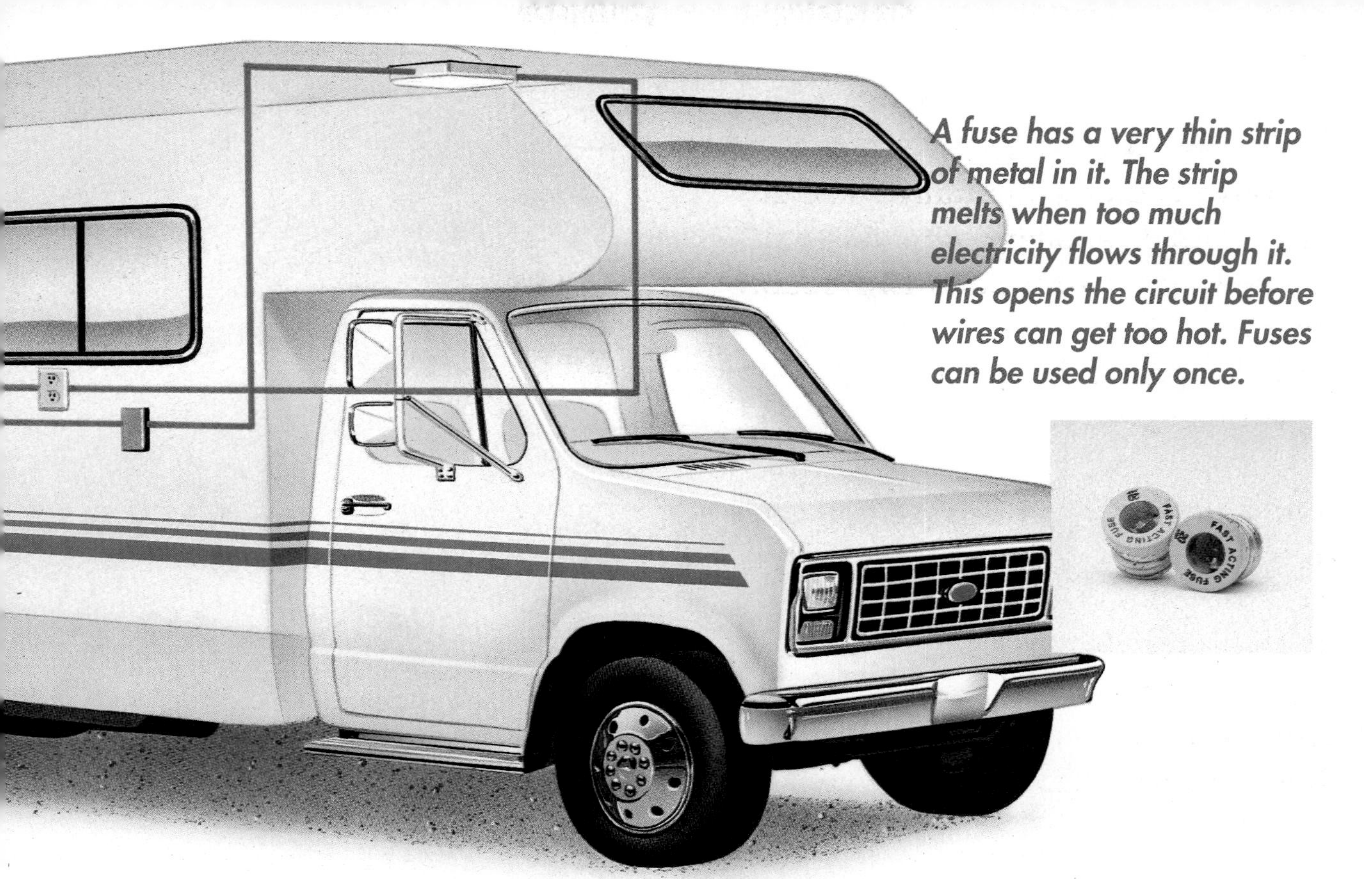

A fuse has a very thin strip of metal in it. The strip melts when too much electricity flows through it. This opens the circuit before wires can get too hot. Fuses can be used only once.

In the Try This Activity, the electric current "skipped" the bulbs. When electrical current skips the rest of a circuit (in this case the light bulbs), it is called a **short circuit.** This can happen if frayed or broken wires touch. Too much current flows through wires when this happens. Very little of the electrical current goes to light a bulb or to run an appliance—it mostly goes to heat up the wire.

Sum It Up

Imagine a world without different kinds of circuits. You couldn't control the lights or appliances in your home. Because we know how electricity moves "down the line" and interacts with different types of circuits, we can control where electricity goes and what it does. This makes part of our lives easier. Understanding simple parallel and series circuits gives you a good idea of how the circuits in your home work.

Critical Thinking

1. Draw a circuit with two bulbs and two switches, one controlling each bulb.
2. Draw a circuit diagram with two switches and one light bulb in which both switches have to be closed for the light bulb to light.
3. If you have a flashlight with one dead D-cell and one good D-cell, will the flashlight work? (HINT: Think of how the D-cells are connected.)

bread
milk
eggs
cereal

Piano lessons
at
3:00 p.m.

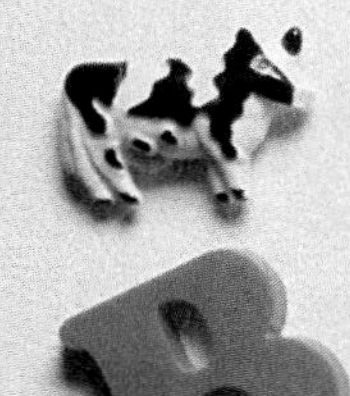

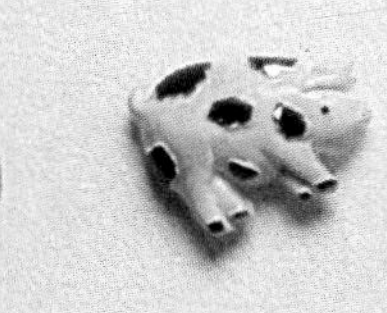

How many of your papers could you tack to a refrigerator with one magnet? Like static electricity, magnets exert a force, and you can observe how this force interacts with objects around you.

Can you think of a time when you moved something without touching it? It sounds impossible, but with magnets it happens. Magnetic marbles and other magnetic toys move each other or stick together.

MAKING IT STICK

You can't see exactly what happens to make the magnets stick together, but you can see the effects. Have you ever wondered how the pull gets from one magnet to another or from the magnet to the refrigerator? The magnets you have used most are probably those on your refrigerator. They hold up pictures, announcements, and bills. Why don't we use magnets to hold pictures on the walls in our homes?

Minds On! Imagine that you have to go through your entire day moving things without touching them. How would you do it? Describe in your ***Activity Log*** on page 20 some of the problems you'd have and how you'd solve them. Then, share your ideas and solutions with a classmate. ●

EXPLORE

Activity!

The Push and Pull of Magnets

You're going to use the way magnets affect other materials to see how they push and pull each other.

What You Need

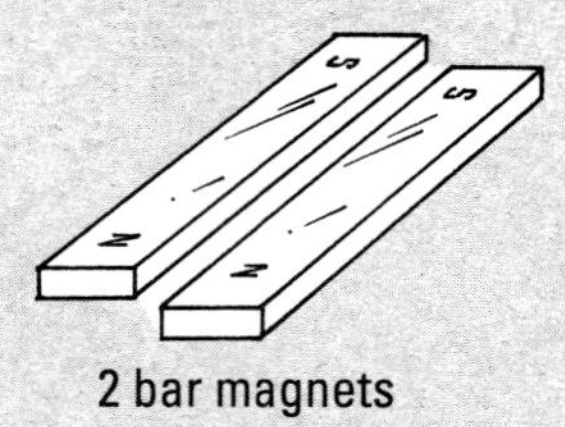
2 bar magnets

piece of white paper

iron filings

horseshoe magnet

safety goggles

plastic cup

3 sealable plastic bags

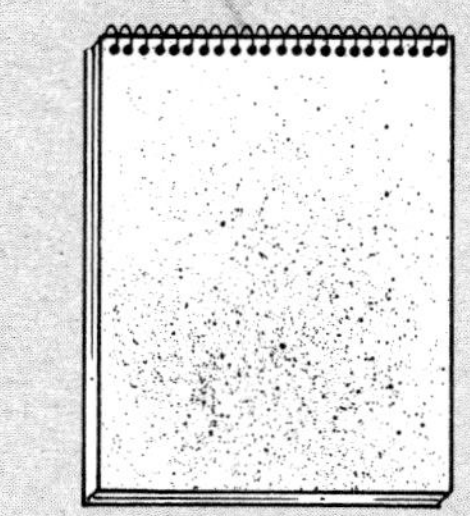
***Activity Log* pages 21-22**

What To Do

1 Place one bar magnet on a desk and try to push it across the surface with another magnet without touching it. Next, try to pull one bar magnet across the desk without touching it. How many combinations can you find that work? Draw the combinations that work. Record your observations in your ***Activity Log.***

2 Put each magnet inside a different plastic bag. Put one magnet flat on your desk. Put the piece of paper on top of it.

Safety! See the *Safety Tip* in step 3.

3 ***Safety Tip:*** Put on your safety goggles. Sprinkle the iron filings on the paper above and around the magnet. Sketch what you see in your ***Activity Log.***

4 Carefully pick up the paper and pour the iron filings into the plastic cup. Set it aside.

5 Put the two bar magnets on your desk with the ends about 2 cm apart. Put the piece of paper on top of the two magnets. Put on your goggles and sprinkle iron filings on the paper above and around the two magnets. Sketch what you see in your ***Activity Log.*** The ends of each magnet are labeled *N* and *S;* look and write down which two ends were facing each other.

6 Repeat steps 4–6 with as many different pairs of ends as possible. Be sure to label which ends are facing each other in your sketches.

7 Repeat steps 3 and 4 for the horseshoe magnet.

What Happened?

1. What difference did you see between the pattern of iron filings with two of the same ends (S and S or N and N) near each other and the pattern with two opposite ends (N and S) near each other?
2. Which ends of the bar magnets pulled toward each other? Where did the iron filings clump closest together?

What Now?

1. Look at your sketches of the patterns the magnets made in the iron filings. How do you think the magnets pushed or pulled each other without touching?
2. Which parts of the magnets would be best for picking up paper clips?

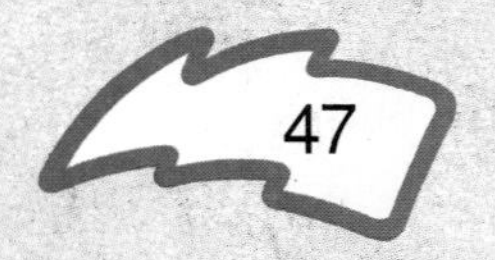

Magnetic Poles

Did you know that magnets are named for a country? About 2,000 years ago, the people in a small part of what is now Greece found rocks that would pull small pieces of iron to them. The area was called *Magnesia* (mag nēz´ shə), and the rocks are still called *magnetite* (mag´ nə tīt´) today. A **magnet** (mag´ nit) is a piece of material or a device that attracts iron-containing materials and some other metals.

More ball bearings are clumped at the poles, where the pull of the magnet is strongest.

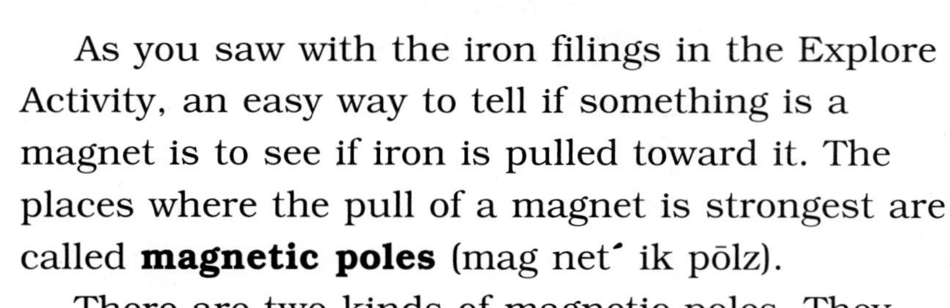

As you saw with the iron filings in the Explore Activity, an easy way to tell if something is a magnet is to see if iron is pulled toward it. The places where the pull of a magnet is strongest are called **magnetic poles** (mag net´ ik pōlz).

There are two kinds of magnetic poles. They are somewhat like the two kinds of electrical charges (+ and –). The same poles (N-N or S-S) push away from each other, while opposite poles (N-S) pull toward each other. Magnetic poles are different from static electricity because once you have a magnet, you don't have to charge it each time to make it work.

Minds On! You have two bars of steel. How could you tell which one was a magnet if the bars weren't labeled and you could use only the two pieces of metal? What tests would you have to do? Describe each test in your ***Activity Log*** on page 23.

TRY THIS

Activity!

What's Attracted?

What You Need

bar magnet inside plastic bag, black paper, salt, *Activity Log* page 24

Is everything attracted to magnets? What would happen if you put salt on a piece of paper over a magnet? Sprinkle some of the salt on the paper over the magnet. What happens? Would sand act the same? How do you know?

The iron filings in the Explore Activity gave you some idea of how magnetic poles act. When opposite poles were near each other (N-S), there were lots of filings pulled between them. When the same poles were near each other (N-N or S-S), the filings were pushed out to the sides. Do the Try This Activity to see how some other materials interact with magnets.

The Magnetic Field

The lines of iron filings aren't actually the magnetic field—they just help you see where it is. The field itself is invisible and surrounds the magnet on all sides.

Look back at the drawings you made in the Explore Activity. The pattern of lines the iron filings made around each shape of magnet was different. But all the patterns had something in common. The lines always curved out from one pole, around, and back to the other pole.

The **magnetic field** (mag net´ ik fēld) refers to the way magnets exert force on iron-containing objects, other magnets, and some special materials. The lines showed you where the magnetic field is. The lines were closest together at the poles because the magnets' pull is strongest there.

The magnets didn't have to touch the filings or other magnets to make them move. Objects with static electricity didn't have to touch to make each other move, either. Electricity and magnetism are forces that can make things move without touching them. This is called *action at a distance.*

You noticed that magnets stick to other magnets. Iron filings are attracted to magnets because each filing becomes a small magnet when it's near a strong magnet. This is why magnets attract iron and things that contain iron, like steel. Each little piece of iron becomes a tiny magnet. You can even make a temporary magnet yourself. Do the Try This Activity on the next page to see how.

TRY THIS **Activity!**

What Makes a Magnet?

What You Need

bar magnet, paper clip, iron filings, safety goggles, *Activity Log* page 25

Make a magnet by pulling a straightened paper clip over a magnet. Pull it at least ten times in the same direction. ***Safety Tip:*** Put on your safety goggles and use your new magnet to pick up some of the iron filings you used earlier. In your ***Activity Log,*** record about how much it picks up. Keep your magnet and test it again tomorrow. How much did the magnet "fade" overnight?

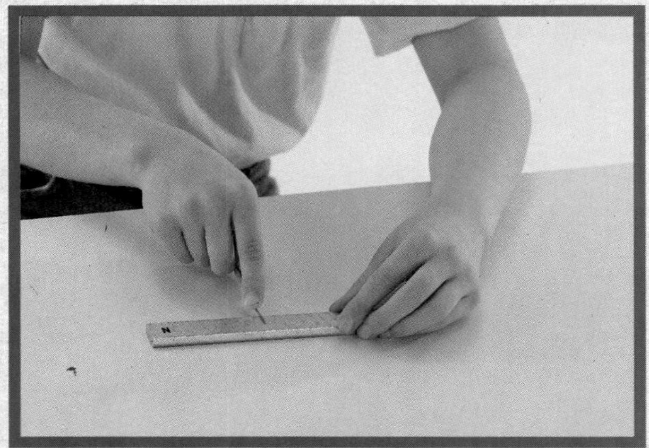

There are other materials that can be magnetized. Ceramic magnets are made from metals that have been combined with oxygen. They are brittle and can break if you drop them on the floor. However, they weigh less than iron magnets. There are even plastic magnets made of many very small magnet pieces embedded in flexible plastic.

How We Use Magnets

Around 300 B.C. a Chinese writer noticed that pieces of magnetite would move if they were hung and allowed to swing freely. After experimenting, it was found that the stones always moved to point the same way. The stones always pointed along a line from north to south. By A.D. 1000 this discovery had spread and Chinese travelers were using pieces of magnetite for navigation. They either suspended pieces of magnetite from string or put them in wooden floats in pans of water.

The idea of using magnetite to find direction eventually reached Europe. In Europe, pieces of magnetite came to be called *lodestones* (lōd′ stōnz), which means "leading stones." Because sailors could now tell direction without depending on the sun and stars, they could navigate on a completely cloudy day or night. Lodestones led them when they needed to know direction. Today, compasses use a small magnet in almost the same way.

What's the Point?

What You Need

string, bar magnet, *Activity Log* page 26

You can recreate what the Chinese discovered. Make a sling for the bar magnet with string. Hang the magnet in the center of your classroom and wait for it to stop moving. In what directions do the poles point? In your ***Activity Log,*** draw your classroom and show how the magnet points. How could you use this device to find your way?

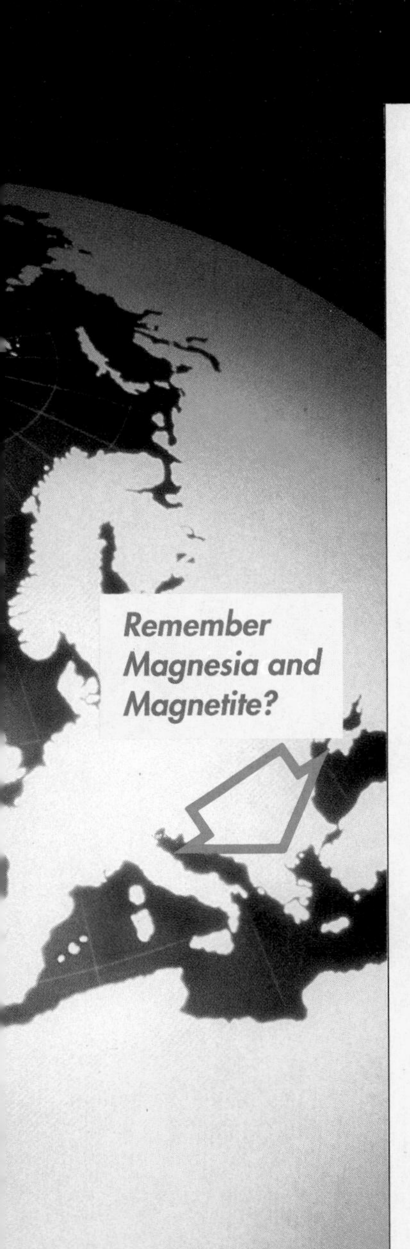

Did you notice which pole of the bar magnet ended up pointing north in the Try This Activity? A magnetic pole is located near Earth's North Pole. The magnetic pole near Earth's South Pole is the opposite pole. Magnets line up to point north and south because of Earth's magnetic poles. The letters *N* and *S* that appear on your bar magnets are short for *North-seeking magnetic pole* and *South-seeking magnetic pole.*

Electricity and Magnets Literature Link

Read *Electricity and Magnets* by Barbara Taylor to find some fun activities that use magnets. You can find out how to make a magnetic window-cleaner, then demonstrate how it works at home.

Sum It Up

Magnetism might never have been discovered if people hadn't observed that magnetite pointed along the line from north to south. If your ancestors didn't come from North America, you might not be here if it weren't for magnets. Early sailors had a very hard time crossing oceans without a way to tell north when the skies were dark and cloudy. We understand more about magnets because of the way they interact with Earth, each other, and objects near them.

Critical Thinking

1. Your compass doesn't work when it's near a bar magnet on your desk. Why is that?
2. What is one possible explanation you could give if you saw two small metal doughnuts threaded on a pencil with one floating a little above the other?
3. How could you compare the strength of two magnets? Why might this be important?

Current Attractions

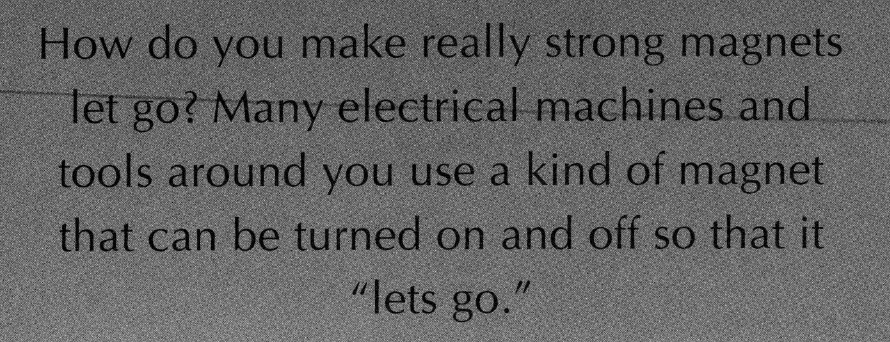

How do you make really strong magnets let go? Many electrical machines and tools around you use a kind of magnet that can be turned on and off so that it "lets go."

If you have ever played with a really powerful magnet, you've probably noticed one problem. Although a strong magnet picks up big objects, it is difficult to get the objects loose again. In the picture, you can see a gigantic magnet. The magnet is so strong that it can pick up an entire car. How will the operator separate the car from the magnet?

Minds On! Imagine some good uses for a big magnet that could be turned on and off. In your ***Activity Log*** on page 27, explain how you would use the magnet. Then, share your ideas with one of your classmates.●

EXPLORE

Activity!

Turning the Field On and Off

You can see how electricity and magnetism are connected by making a magnet using electric current. You'll see how it can be turned on and off and find out what controls its strength.

What You Need

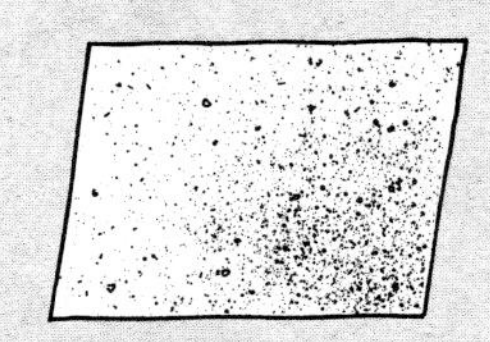

small piece of sandpaper

2 D-cells

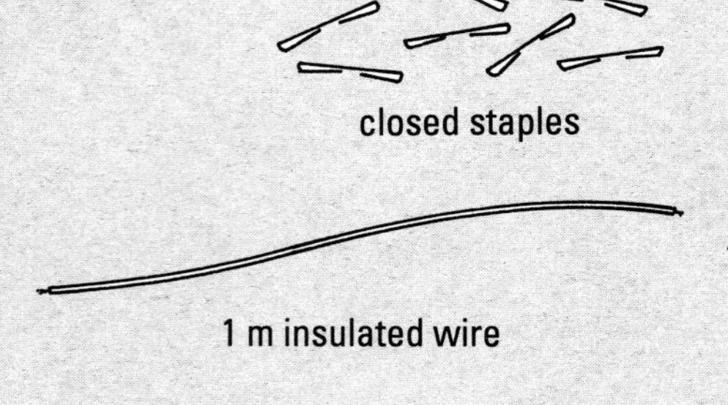

closed staples

1 m insulated wire

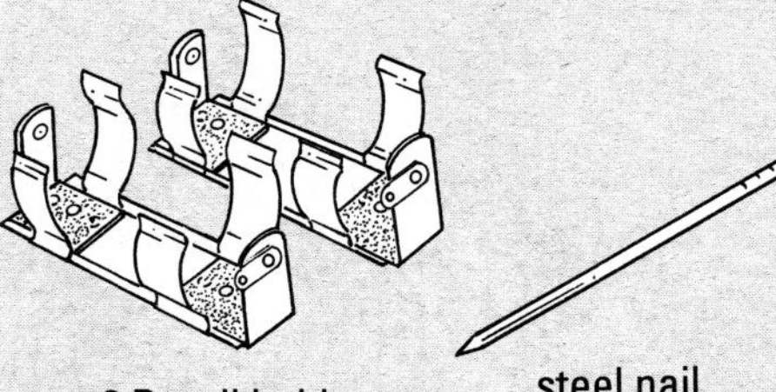

2 D-cell holders

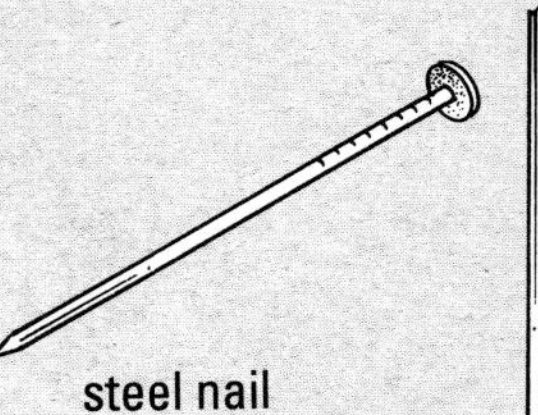

steel nail

***Activity Log* pages 28-29**

What To Do

1 Touch the nail to a pile of staples and see if any of them will stick to the nail. Record your observations in your ***Activity Log.***

2 Leaving about 20 cm of extra wire at the beginning and end, wrap 15 loops of wire around the nail. Be careful not to overlap any loops.

3 Since the enamel coating on the wire is an insulator, use the sandpaper to clean the enamel coating off the last 1–2 cm at each end.

4 Connect one end of the wire to the D-cell holder. Have one partner hold the other end of the wire to the other end of the D-cell holder while you carefully move the nail into the pile of staples.

5 Pick up as many staples as you can. Move the nail away from the pile, and then take the end of the wire away from the D-cell holder.

6 Count the number of staples you picked up. Record that number on the chart in your ***Activity Log.***

7 Predict what you have to do to make the nail pick up more staples. Test your prediction by using the materials you have to change your setup and repeating steps 4–6.

What Happened?

1. What happened when you placed the nail in the staples when the circuit was closed?
2. What happened when the circuit was opened (when you took one end of the wire away from the D-cell)?
3. What did you change to make the nail pick up more staples?

What Now?

1. Why was the nail able to pick up the staples?
2. What could you do to make the nail pick up even more staples?

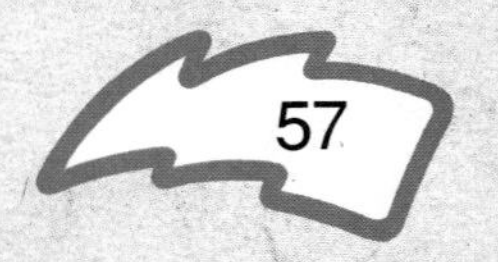

What Makes an Electromagnet?

The nail wasn't much of a magnet before you wrapped the wire around it. It didn't act like a strong magnet at all until you closed the circuit and let electrical current flow through the wire. When you opened the circuit, the nail lost most of its magnetic properties.

TRY THIS

Activity!

What Makes It Stick?

What You Need

magnet from the Explore Activity, iron filings, safety goggles, *Activity Log* page 30

Take the coil of wire off the nail. Connect it to a D-cell and try to pick up some staples using just the coil. How well does your magnet work? Try the iron filings test with and without the nail in the coil. ***Safety Tip:*** Don't forget to wear your safety goggles! What happens? In your ***Activity Log,*** draw the patterns you saw. What does this tell you about what makes the magnet work well?

When an electric current moves through a wire, it makes the wire a magnet. By itself the wire isn't a very strong magnet. You don't even notice the magnetic properties of the wires in your home. When a wire is coiled around and around, it concentrates the electric current around one place. This kind of magnet is stronger, but it won't pick up much material.

Putting a piece of iron or steel inside the coil makes the magnet stronger. While the electric current is flowing, the iron or steel will act just like a magnet. It can pick up small iron objects and it has poles, just like a magnet. When the current is turned off, the nail stops acting like a magnet and drops any objects it was holding. This arrangement—a core of iron or steel wrapped in a coil of electric current—is called an **electromagnet** (i lek´ trō mag´ nit).

The compasses would all point in the same direction if it weren't for the wire. The current flowing in the wire is producing a magnetic field. The magnetic field is making all the compasses point toward the wire.

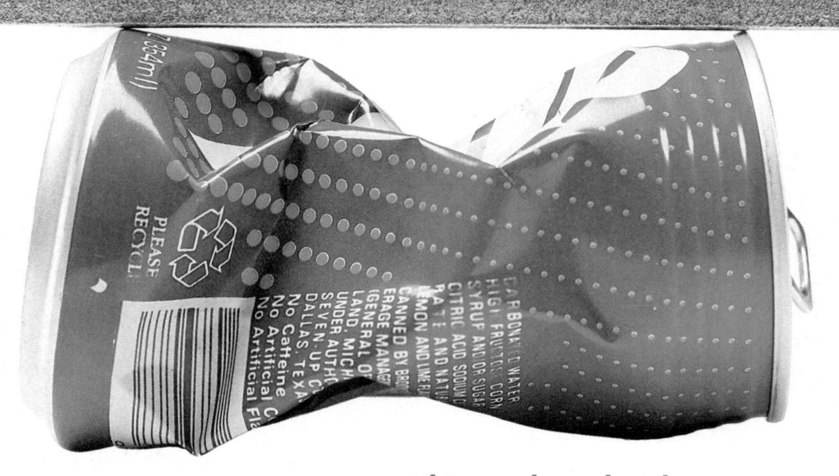

This steel can has been separated from the aluminum cans below by a magnet.

Magnet Words

Language Arts Link

Like how the object works, the word *electromagnet* is made by sticking two words together. Make a list of all the words you know that have *electro-*, *electricity*, or *magnet* in them. Do the words all describe things that use both electricity and magnetism? Make up some words that have both words in them.

The aluminum cans aren't pulled up to the magnet. This fact can be used to separate material wanted for recycling from unwanted material.

How We Use Electromagnets

Electromagnets come in all sizes, from little ones that make doorbells ring by pulling a striker rapidly forward, to enormous magnets that stick to and pick up entire cars.

Some soda pop cans are made of steel and some of aluminum. Recycling centers want to separate the aluminum cans from the steel ones. Since all steel has iron in it, electromagnets are used to pull the steel cans out from among the aluminum ones.

Electromagnets control what you hear and what you see, too. In a cassette player, the shiny rounded metal parts with black parts on the inside are the magnets. You can't see inside a VCR easily because of the way it's built, but it has magnets, too. TV picture tubes and stereo speakers contain powerful electromagnets. Based on what you know about electromagnets and recording tape, why isn't it a good idea to keep VCR tapes right on top of TVs and speakers? Do the Try This Activity to see how cassette players and VCRs use electromagnets to record and play information.

TRY THIS Activity!

Writing on Tape

What You Need

cassette player, blank cassette tape, pencil, electromagnet from the Explore Activity, *Activity Log* page 31

Connect the electromagnet to the D-cell. Run the point of the nail near the tape as shown. Put the pencil through one of the holes in the center of the cassette and advance the tape a few centimeters. Run the point of the nail near the tape again. Do this 8 to 10 times. Rewind and play the cassette. Describe in your ***Activity Log*** what you hear. What does this tell you about how recorders put sound on tape?

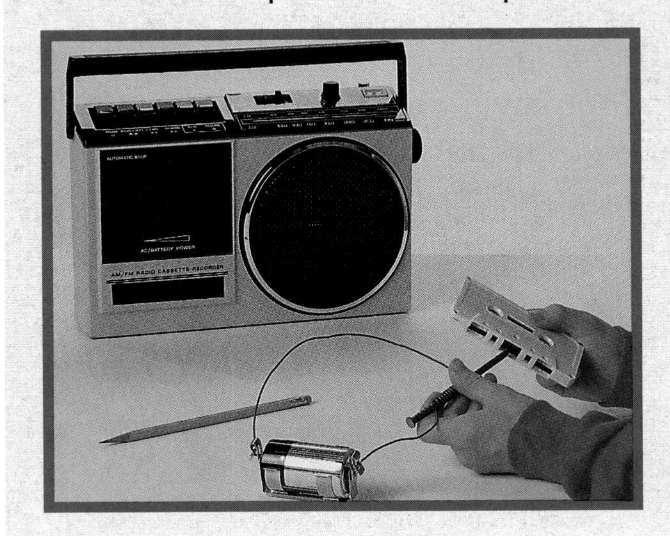

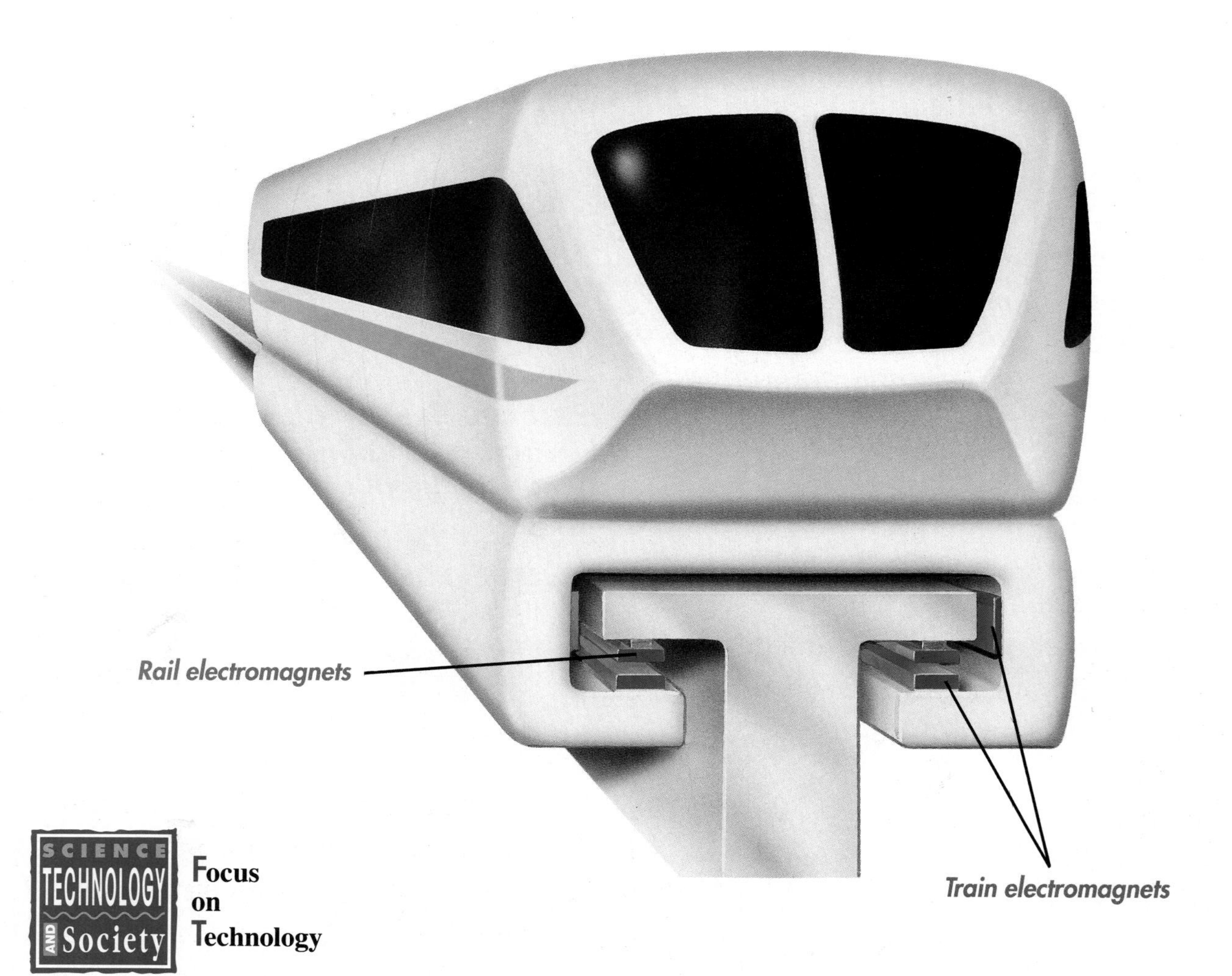

SCIENCE TECHNOLOGY AND Society

Focus on Technology

Mag-lev Trains

Many big electromagnets are used to lift and drive trains called *mag-lev trains. Mag* is short for *magnetic. Lev* is short for *levitation.* The magnets actually lift the train above the rails. Can you guess what *levitation* means? Use a dictionary to check your guess.

Mag-lev trains have no wheels. Instead, they use magnets to run above very special rails. A mag-lev train can go as fast as 500 kilometers (about 300 miles) per hour. It moves fast and smoothly because none of it is actually touching the rail. There is no friction between wheels and rails. There are no wheels to carry the bumps in the rail to the passengers. Because mag-lev trains don't have to use energy to overcome friction, they may eventually be more efficient than regular trains. Right now, some are in use in Japan and Germany. However, they are still being designed, tested, and improved.

Giant electromagnets are fixed on the bottom of the train and ride below the rail. When the train is started up, the magnets are pulled toward the rail and lift the train off the ground. The strength of the magnets is adjusted so that they do not actually attach the train to the rail, but make the train float just above the rail.

Sum It Up

The electromagnet is a tool that we were able to develop because we understand how electricity and magnetism work. Electromagnets help us use electrical energy to make things move. Like current electricity and circuits, electromagnets are part of small things we use every day, such as cassette players and VCRs. They are also part of enormous machines that move, like mag-lev trains and the magnets that lift cars in junkyards.

All these applications of electromagnets came about because we can change parts of a system to make electromagnets as strong as we need them to be and turn them on and off when we want to.

Critical Thinking

1. What would happen if you made an electromagnet using a bar magnet in place of the nail?
2. Does lightning generate a magnetic field? How do you know?
3. You know a cassette recorder "writes" on tape with magnets. What has to be true about the tape for this to work?

Power Up!

What would you do if the lights went out? Would you think about making electricity yourself?

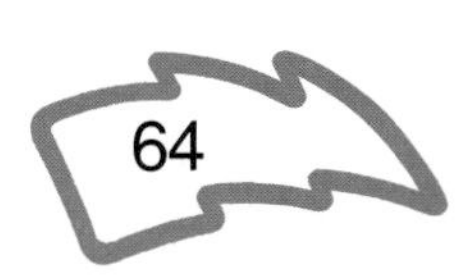

Imagine that you are camping in the woods where there are no power lines and no D-cells. Now, think of all the things that you like to use that require electricity—either D-cells or household current. What could you do to make electricity to run those things?

Electricity wouldn't be much use to us if we couldn't have it when and where we want it. You might like to be able to play a video game in the car. Sometimes you want to use the hair dryer in the bathroom. In either case, you need a source of electricity. D-cells are a portable source of electricity. Wherever you go, you can take them along.

Where does household current come from, though? It comes into the home on wires and is not at all portable. In fact, you sometimes have to look around a room for an outlet close enough to the place where you want to use something.

Minds On! Remember all the things you've done with electricity and wires. Can you think of an easy way to tell if electric current is flowing in a wire? Can you think of a way that doesn't require using light bulbs or disconnecting the wire? ●

EXPLORE

Activity!

What Makes Electric Current?

It's easier than you think to make electric current. You need to make a current detector first, because the electric current probably won't be enough to light a bulb. Then, you'll see if you can detect a current you make yourself.

What You Need

D-cell

D-cell holder

transparent tape

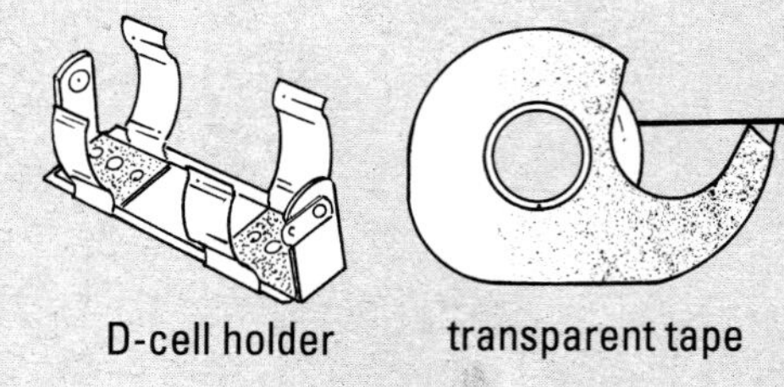

1 m insulated wire with ends stripped

compass

bar magnet

cardboard tube

3 m insulated wire with ends stripped

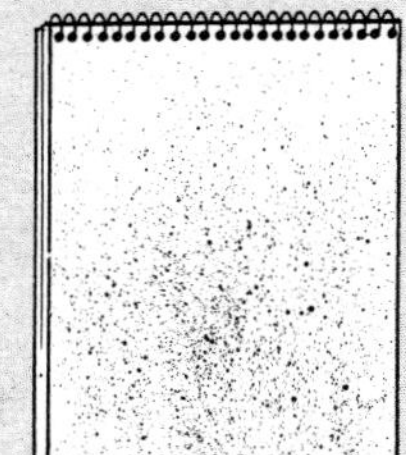

***Activity Log* pages 32-33**

What To Do

1 Wrap the compass with the 1-m wire. You need 8 to 10 loops. The loops should go from north to south. Leave about 20 cm of extra wire at each end.

2 Tape the compass and wire loops to your desk so that the needle points north and lines up with the wire loops. Connect one end of the wire to the D-cell and touch the other briefly to the other end of the D-cell. What happens? Disconnect the D-cell. What happens? Record your observations in your ***Activity Log.*** This device is your current detector.

3 Wrap the 3-m wire around the cardboard tube. Make sure not to overlap any loops. Make the loops as close together as possible and leave at least 50 cm of extra wire at each end.

4 Connect your current detector to the cardboard tube wire, one wire end to one wire end. You can use an empty D-cell holder to hold the wires together.

5 Make sure the tube is as far away from the current detector as possible so that the magnet won't directly affect the compass. Insert the bar magnet into the cardboard tube while one of your partners watches the compass. Record your observations in your ***Activity Log.*** Predict what will happen if you insert the other pole of the magnet.

6 Test your prediction.

What Happened?

1. What happened to the compass when current ran through the wire?
2. What happened when you pushed the magnet into the cardboard tube?

What Now?

1. What did the loops of wire around the compass needle form when a current flowed through them?
2. In what way is the compass needle's movement evidence of a current in the wire?
3. What produced a current in the wire when you moved the magnet?

What Makes the Electricity You Use?

Household Current

You knew that an electrical current flowing in a coil of wire could make a magnet. Now you know that a moving magnet in a coil of wire can produce an electrical current. When you moved the bar magnet in the coil of wire in the Explore Activity, it produced weak electric current. You detected the current by its effect on the compass.

The kind of current you use in your home is like the current made by the magnet. It's called **alternating current** (ôl tər nāt´ ing kûr´ ənt), or *AC* for short, because it changes. What changes is the direction in which the current flows.

Moving the magnet into the cardboard and wire tube made the compass needle swing one way. Pulling it out made it swing the other way. The reason you don't see the lights in your house flicker when the AC changes direction is that it changes many times each second. The change is so quick that you don't notice it.

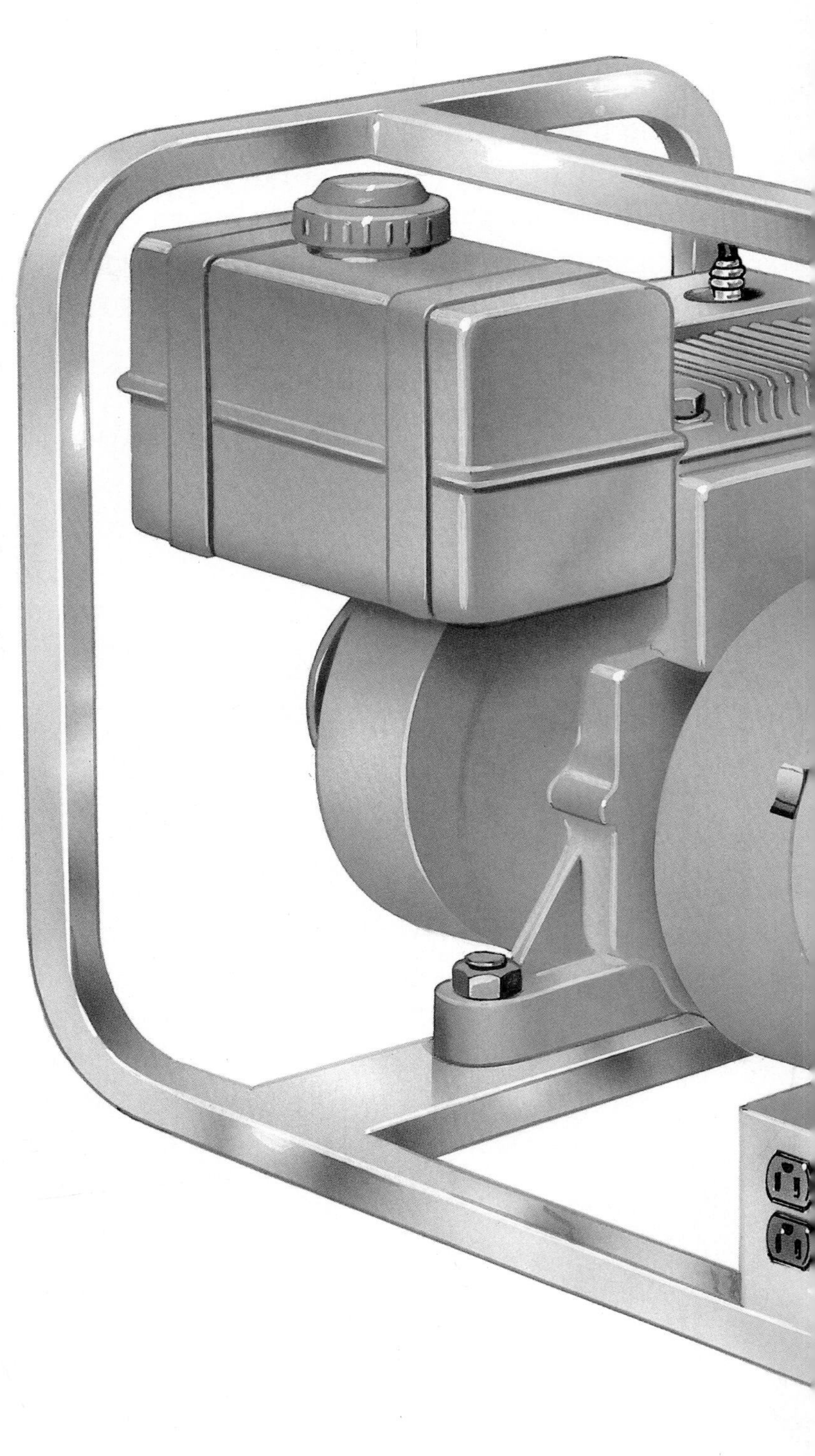

Small, portable generators are usually turned by gasoline engines. They still have the same basic parts as the bigger generators that make electricity for your home. There is a stationary magnet, or stator, and a coil of wire, or rotor, that turns around the stator. The rotor can be inside or outside the stator.

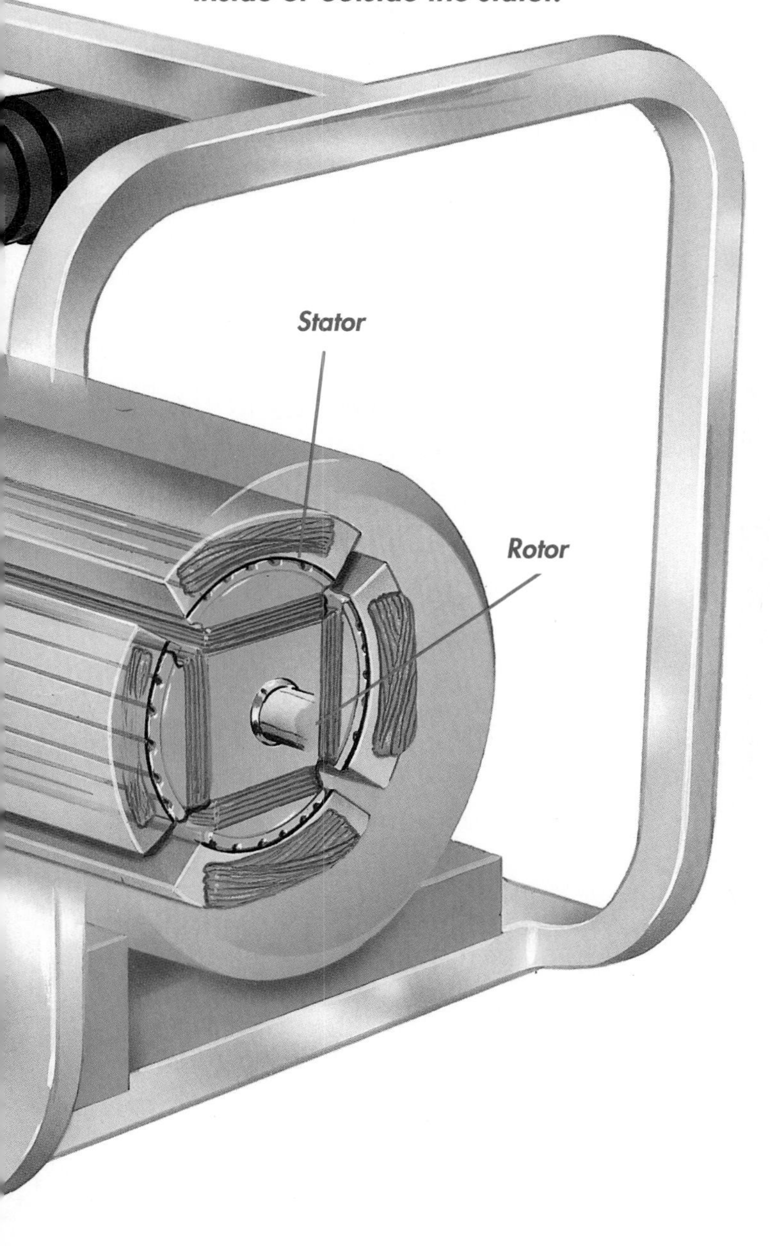

We call the device that makes alternating current a **generator** (jen´ ə rā´ tər). The bigger generators that make the AC electricity you use aren't quite like your simple wire coil and magnet. They use the same basic parts, coils and magnets, but they use them differently. In bigger generators, a coil of wire spins between two powerful magnets. A magnet doesn't move in and out of the coil of wire. It is much easier to make a generator that turns than one with a magnet going in and out.

The energy that makes a generator turn can come from many sources: water falling down from a dam or expanding steam heated by a coal or nuclear power plant. The energy of the moving water or moving steam is transferred to the turning motion of a big, fan-like device called a *turbine*. The turbine is connected to a generator and makes it turn.

TRY THIS **Activity!**

Wet Current

Safety Tip: Vinegar is an acid and can damage eyes and sensitive tissue.

What You Need

half cupful of distilled vinegar, thin copper strip, thin zinc strip, plastic cup, current detector from the Explore Activity, safety goggles, *Activity Log* page 34

Put your goggles on. Pour the vinegar into the cup. Connect one strip to each end of the current detector. Watch the current detector while your partner puts one strip in, then holds both strips in the vinegar. Describe in your ***Activity Log*** how the result is similar to and different from the result of the Explore Activity.

D-Cell Current

Generators aren't the only possible source of electric current. The D-cells you've used till now produce a current with chemicals. They contain a paste that conducts. The paste separates two materials, one of which has extra available electrons. When you put a D-cell into a circuit, the extra electrons flow through the circuit.

There's a difference between D-cells and batteries. You probably know that cells are small parts of living things. **Electric cells** (i lekʹ trik selz) are small parts of batteries. You made an electric cell with the metal strips and vinegar. A **battery** (batʹ ə rē) is a device made up of several cells connected in series. A 6-volt battery is really four 1.5-volt cells hooked together.

The inside of a D-cell

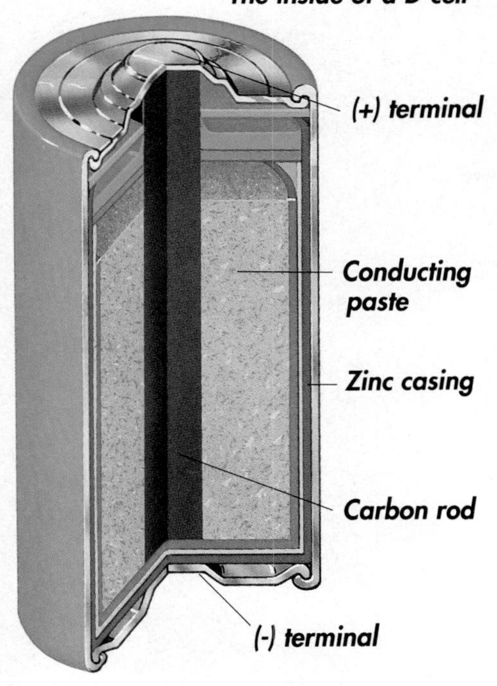

The paste inside D-cells and batteries contains a chemical that can burn you. Never try to cut or break one open to play with what's inside!

Because the current from D-cells and batteries flows in only one direction, it's called direct current (di rekt´ kur´ ent), or DC for short. The batteries in cars produce direct current. All the small, portable home appliances and toys you've seen that use batteries run on DC. We use DC because it's portable and because it can be produced cheaply and easily using cells and batteries.

Batteries—Small But Powerful!

Focus on Technology

Batteries are a convenient, easy-to-use source of electric energy for familiar items such as toys, flashlights, and even TV sets. Batteries of various sizes and strengths can be purchased almost anywhere. But batteries have not always been so easily obtained or used.

One of the first batteries was developed by an Italian scientist Count Allesandro Volta in the late 1790's. It consisted of pairs of silver and zinc disks. Batteries have evolved from Volta's first crude invention. We now have submarine batteries that weigh almost 1 metric ton. We also have tiny lithium batteries that weigh 1.4 grams.

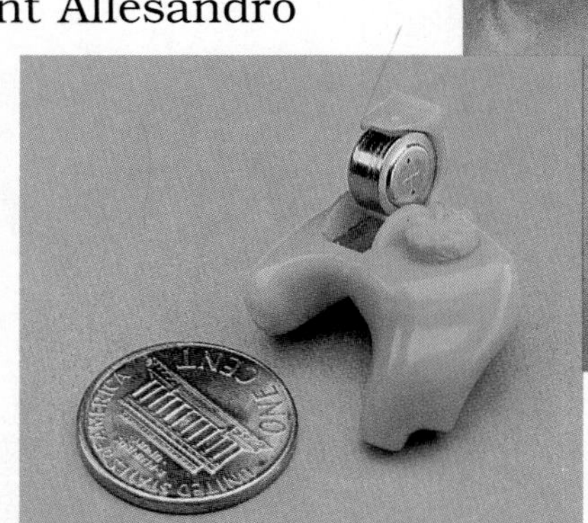

A button battery, used in hearing aids, produces voltages higher than any other single cell.

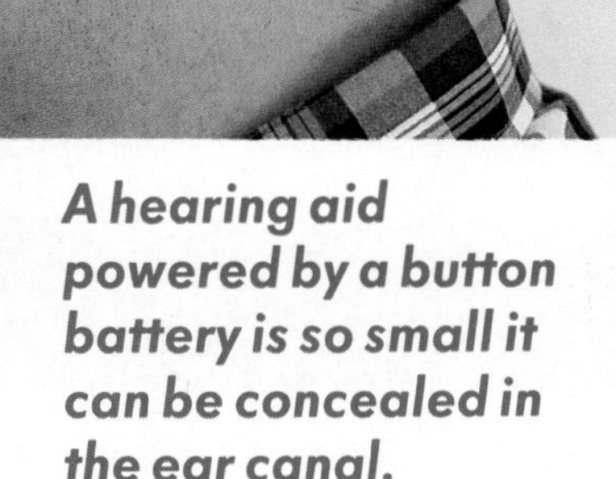

A hearing aid powered by a button battery is so small it can be concealed in the ear canal.

Conserving Energy

Remember what happened in New York City when the lights went out? You know that many people could not live the way they do without electricity. However, producing electrical energy costs money. Although electricity itself is environmentally clean, producing it can pollute the environment. Most plants that generate electricity burn coal or oil, or use nuclear fuels that do pollute the environment. Engineers and scientists are developing ways to make these generating plants work more efficiently and produce less pollution.

Engineers also help you use less energy by designing more efficient machines. You can help use less electricity in some simple ways. The following suggestions will help you conserve energy even if you don't have many electrical appliances.

- Don't waste hot water. It takes a lot of energy to heat water.
- Keep doors and windows closed if a heater or an air conditioner is on. Heating or cooling the outdoors would take forever and would use up lots of electricity!
- Turn off lights and other electrical devices when they are not being used. Don't leave the television on if you aren't watching it.

Sharing Saving Ideas   Music/Art Link

Make a poster that illustrates one of the energy-saving suggestions above or another suggestion you think of. When all the posters are finished, hang them in one section of your classroom.

Sum It Up

Think back to your campsite. How would you get electricity to your campsite? If you could find a way to make a magnet interact with a coil of wire, you could make some electrical current. But you'd still have to work hard. Can you think of a way to use something at your campsite besides you to make a generator turn?

At home the electricity you use costs money, not time and energy. Although electricity seems very clean when you use it, making it usually produces pollution of some kind. It's always wisest to use only as much electricity as we need.

Critical Thinking

1. Earth has a magnetic field. How could you use that field to generate a small electric current?

2. Why do we sometimes use 9-volt batteries instead of several D-cells?

3. Why do you think we don't use small, hand-powered generators for household appliances?

History of Using Electrical Energy

Just over 100 years ago, the first electric power plants were opened—one in San Francisco and another in New York City. About the same time, Thomas A. Edison is credited with inventing the light bulb. Ten years later, the Irish physicist G. Johnstone Stoney described electric current as a flow of tiny particles that he called *electrons.*

One hundred years is only about ten times your age. During the past 100 years, the entire world has been electrified. There is probably not a single nation on Earth that does not provide electricity for at least some of its people. Many countries supply electricity for most of their population. Imagine the number of power plants that have been built, the miles of wire that have been strung, and the number of buildings that have been wired in such a short time.

Think of the uses for electricity that have been discovered in the last 100 years: lights, telephones, telegraphs, radios, clocks, washing machines, dryers, refrigerators, sewing machines, movie projectors, televisions, video cassette recorders, phonographs, CD players, audiotape recorders, automatic teller machines, robots, typewriters, electronic printers, calculators, and computers.

Skyglow of the United States

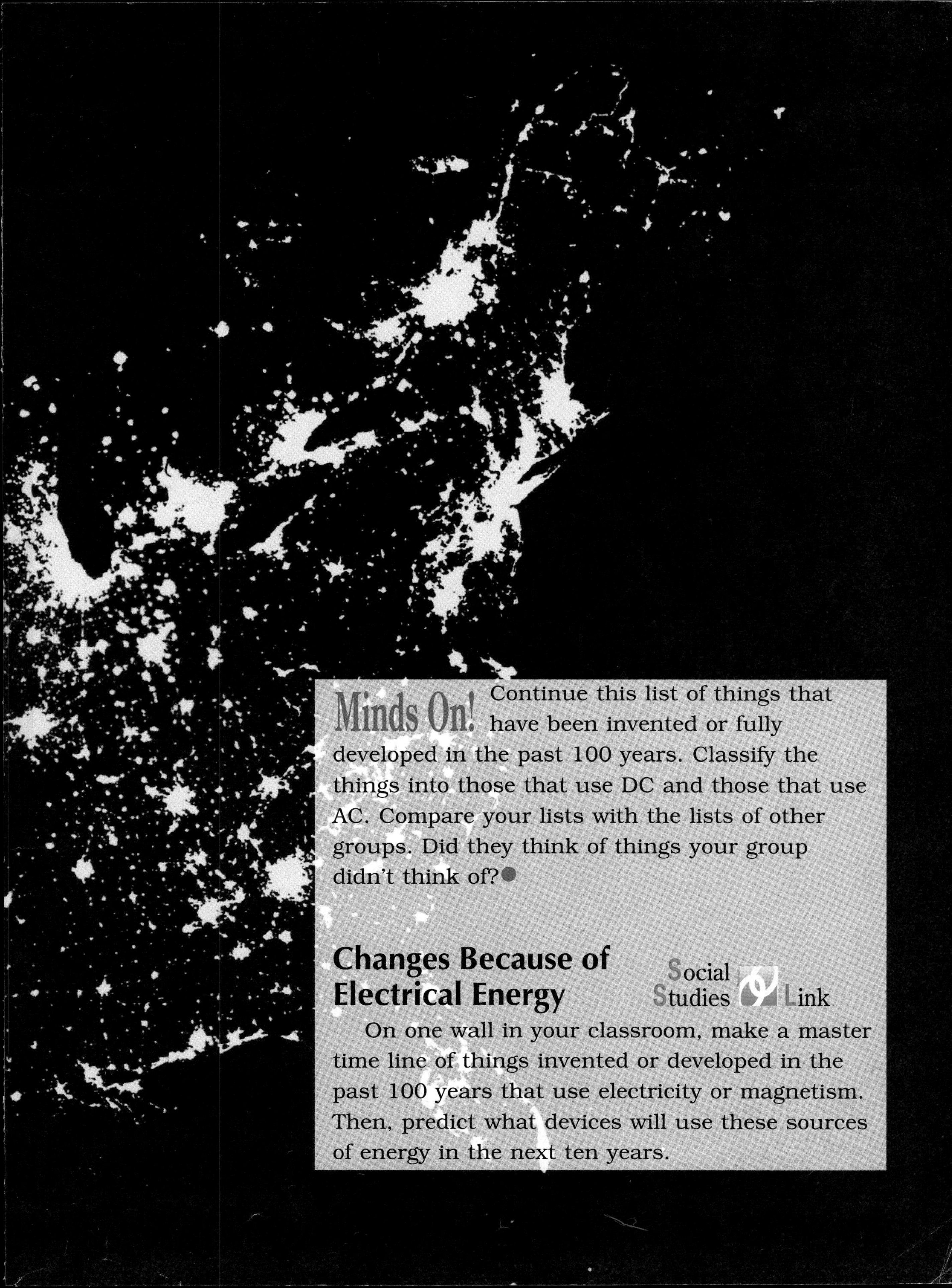

Minds On! Continue this list of things that have been invented or fully developed in the past 100 years. Classify the things into those that use DC and those that use AC. Compare your lists with the lists of other groups. Did they think of things your group didn't think of?

Changes Because of Electrical Energy

Social Studies Link

On one wall in your classroom, make a master time line of things invented or developed in the past 100 years that use electricity or magnetism. Then, predict what devices will use these sources of energy in the next ten years.

GLOSSARY

Use the pronunciation key below to help you decode, or read, the pronunciations.

Pronunciation Key

a	at, bad	d	dear, soda, bad
ā	ape, pain, day, break	f	five, defend, leaf, off, cough, elephant
ä	father, car, heart	g	game, ago, fog, egg
âr	care, pair, bear, their, where	h	hat, ahead
e	end, pet, said, heaven, friend	hw	white, whether, which
ē	equal, me, feet, team, piece, key	j	joke, enjoy, gem, page, edge
i	it, big, English, hymn	k	kite, bakery, seek, tack, cat
ī	ice, fine, lie, my	l	lid, sailor, feel, ball, allow
îr	ear, deer, here, pierce	m	man, family, dream
o	odd, hot, watch	n	not, final, pan, knife
ō	old, oat, toe, low	ng	long, singer, pink
ô	coffee, all, taught, law, fought	p	pail, repair, soap, happy
ôr	order, fork, horse, story, pour	r	ride, parent, wear, more, marry
oi	oil, toy	s	sit, aside, pets, cent, pass
ou	out, now	sh	shoe, washer, fish mission, nation
u	up, mud, love, double	t	tag, pretend, fat, button, dressed
ū	use, mule, cue, feud, few	th	thin, panther, both
ü	rule, true, food	th̲	this, mother, smooth
u̇	put, wood, should	v	very, favor, wave
ûr	burn, hurry, term, bird, word,courage	w	wet, weather, reward
ə	about, taken, pencil, lemon, circus	y	yes, onion
b	bat, above, job	z	zoo, lazy, jazz, rose, dogs, houses
ch	chin, such, match	zh	vision, treasure, seizure

alternating current (ôl´tər nāt´ing kûr´ənt) an electric current that regularly changes the direction it flows. In the U.S. it changes 60 times per second

atom (at´əm) the smallest piece of an element that still has the properties of that element

battery (bat´ə rē) more than one electric cell connected in series or in parallel

circuit (sûr´kit) a closed path for electricity to flow through, usually, but not necessarily a wire made of a conductor

circuit breaker (sûr´kit brāk´ər) a switch-like device that protects electrical circuits from excess current

circuit diagram (sûr´kit dī´ə gram) an engineering plan that uses lines and symbols to represent electrical circuits

closed circuit (klōzd sûr´kit) a circuit with no breaks or interruptions

conductor (kən duk´tər) a material that electricity flows through with relative ease. Good conductors have low resistance.

current electricity (kûr´ənt i lek tris´i tē) moving electric charges

direct current (di rekt´ kûr´ənt) the electric current provided by batteries and electric cells. Direct current only flows in one direction

electric cell (i lek´trik sel) a single source of direct current that changes chemical potential energy to electrical energy

electricity (i lek tris´i tē) anything involving electric charges, but usually refers to static, direct current, and alternating current electricity

electromagnet (i lek´trō mag´nit) a magnet produced by an electric current, usually a core of iron wrapped in a conductor

electron (i lek´tron) a particle of matter much smaller than an atom that carries a negative charge

fuse (fūz) a device to protect electrical circuits from excess current, usually a thin piece of metal mounted in a threaded, glass-topped case

generator (jen´ə rā´tər) a device that uses a moving magnetic field and a coiled conductor to produce alternating electric current

grounding (ground´ing) discharging an electric charge to Earth

insulator (in´sə lā´ tər) a material with a very high resistance to electricity

magnet (mag´nit) a thing or device that attracts other magnets, iron-containing materials, and other metals such as nickel

magnetic field (mag net´ik fēld) the invisible field that carries magnetic force to other objects

magnetic pole (mag net´ik pōl) the area of a magnet where its attraction is strongest

open circuit (ō´pən sûr´kit) a circuit with a break in it

parallel circuit (par´ə lel´sûr´kit) all or part of an electrical circuit that electric current divides to flow through

proton (prō´ton) a particle of matter smaller than an atom that carries a positive charge

rechargeable battery (rē charj´ ə bəl) a battery that can have energy added by household current and a special recharger. Once most batteries are used up, they can't be recharged.

resistance (ri zis´təns) a measure of how well a given material will conduct electricity. Good conductors have little resistance. Light bulbs have a higher resistance. Insulators have a very high resistance.

series circuit (sîr´ēz sûr´kit) all or part of an electrical circuit where electric current is undivided and flows through one circuit part after another

short circuit (shôrt sûr´kit) a dangerous condition in which electric current flows through a path of very little resistance and bypasses all or some of the resistors in a circuit

static electricity (stat´ik i lek tris´i tē) an excess of non-moving electric charge in one place, caused by an excess or a lack of electrons

switch (swich) a device to open or close an electrical circuit

volt (vōlt) a measure of the "push" provided by an electric current source

INDEX

CREDITS

Photo Credits:
Cover, The Image Bank/Lionel Brown; David Rowley/Natural Selection; **2, 3,** Graham Finlayson/Tony Stone Worldwide/Chicago Ltd.; **3,** ©Studiohio; **6, 7,** David Rowley/Natural Selection; **8,** (b) The Bettman Archive, (t) ©Doug Martin; **9,** (R) ©Lowell Georgia/Photo Researchers, (l) Tom Till/INT'L Stock; **10, 11,** ©Studiohio; **12, 13,** RM International Photography/Richard Haynes; **14, 15,** ©Studiohio; **18, 19,** ©KS Studios/1991; **20,** (l) Perry Conway/Tom Stack & Associates, (r) Superstock; **22,** (inset) UPI/The Bettman Archive; **22, 23,** ©KS Studios/1991; **24, 25,** ©Studiohio; **26,** ©KS Studios/1991; **27,** (B) ©KS Studios, (T) ©Studiohio; **28, 29,** ©KS Studios/1991; **28,** (bl) ©Studiohio; **30,** ©Studiohio; **31,** ©Historical Pictures; **32, 33,** (t) ©George D. Lepp/Comstock Inc., (b) E.P.I. Nancy Adams/Tom Stack & Associates; **34, 35,** Paul Meredith/Tony Stone Worldwide/Chicago Ltd.; **36,** ©Studiohio; **40, 41,** ©Fredrik Marsh; **42,** (t) Bruce Coleman Inc./Norman Owen Tomalin, (B) ©Studiohio; **43,** ©George C. Anderson; **44, 45,** ©Fredrik Marsh; **46, 47, 48, 49, 50, 51,** ©KS Studios/1991; **52, 53,** IMTEK Imagineering/Masterfile; **54, 55,** ©Brent Jones; **56, 57, 58, 59,** ©Studiohio; **60,** ©KS Studios/1991; **61,** ©Studiohio; **63,** The Image Works; **64, 65,** Tom Bean/ALLSTOCK; **66, 67, 70,** ©Studiohio; **72, 73,** Grahm Finlayson/Tony Stone Worldwide/Chicago Ltd.; **73, 74,** Hank Brandli.

Illustration Credits:
12, Michael Compton; **14, 24, 36, 46, 56, 66,** Bob Giuliani; **16,** Bill Singleton; **17, 38, 39,** Henry Hill; **42, 43,** Ian Greathead; **62, 68, 69, 71,** James Shough